A Beginner's Guide

Model Steam
Locomotives

An entertaining and informative manual to steer you through the
pleasures and pitfalls of building and running a model loco.

A Beginner's Guide To

Model Steam Locomotives

TIM COLES

Patrick Stephens, Wellingborough

First published in 1986

British Library Cataloguing in Publication Data

Coles, Tim
A beginner's guide to model steam locomotives.
1. Locomotive boilers—Models
I. Title
625.1′9   TJ630

ISBN 0-85059-761-7

*Patrick Stephens Limited is part of the
Thorsons Publishing Group*

Printed and bound in Great Britain.

Contents

Foreword

By Stan Bray

The construction of model steam locomotives has long been one of the leading interests in the field of model engineering. All too often though, models are started but never finished because the beginner encounters too many problems on the way. With guidance this need not be so. Previous books on the subject have covered one particular type of locomotive or have been general books with information on various aspects of this branch of the hobby. Tim Coles covers the whole construction in a logical sequence giving the beginner and experienced model engineer alike an idea of what is involved and how to go about it.

There is no doubt that a model steam locomotive makes a very fine model and completion of one is an achievement for the person concerned. Apart from the beauty of such a model, there is the thrill of being able to actually drive it, and of course the smell of a locomotive is something quite unique and this too is captured by such a model.

Even if you are not proposing to build a model, the book makes fascinating reading, and who does not like to dream of what he or she might make one day? This book will, I am sure, change many of those dreamers into modellers and will encourage those who are beginning to get somewhat frustrated with their modelling, giving them new enthusiasm and the urge to press on and complete their model.

Acknowledgements

I should very much like to thank the many people who have helped me in the production of this book. Without the enthusiastic aid of the following folk, my task would have been very much greater.

Liz Scott converted my longhand scribble into beautiful typescript, while Diane Thurley converted my original drawings into a form suitable for publication. Alec Farmer of A. J. Reeves and Co Ltd generously supplied castings for me to machine and photograph. Denis Bishop and Edgar Richardson of Bishop Richardson Model Engineers allowed me to spend a pleasant morning at their small factory in Birmingham and to photograph them at their work.

Various machine tool companies supplied excellent pictures of their wares and Johnson Matthey Metals Ltd supplied much helpful information about their products. Thanks to Don Coventry of the Southern Federation of Model Engineering Societies for arranging permission to reproduce their list of member clubs and to Margaret Moore for the same service in regard to the Northern Association of Model Engineers. Stan Bray provided encouragement and advice which was extremely welcome. Beryl Hanvey of Patrick Advertising Associates deserves a mention for organizing some of the loco photographs and of course, thanks to all those who let me photograph their engines. Dunstan and Ged Holt and Bill Mawson were outstanding here.

Finally and most important of all, thanks to my wife Caroline who not only did a fair slice of typing and most of the proof reading, but also gave me the time off from my domestic duties to work on the book. As the daughter of a model engineer, she knew what to expect!

All the photographs and drawings in this book are by the author unless otherwise credited.

Tim Coles
Elsworth, Cambridge, August 1985

Tim Coles in his workshop. (Caroline Coles.)

Chapter 1

Start here

There can be little doubt that railways in general and steam locomotives in particular have a powerful fascination for people of all ages and backgrounds. This interest finds an outlet in many forms. At one end of the scale there are fifty or so full size steam railways now carrying fare-paying passengers in this country. These are manned almost entirely by enthusiastic amateurs who give freely of their time and energy to see locomotives operating again in the way they were designed to do. However, steam engines are just as demanding of heavy labour in the way of maintenance as they ever were, and only a few lucky people will get the chance to drive a locomotive.

Going to the other end of things, many folk find satisfaction in railway modelling on a small scale. This can be as tiny as the incredible Z-gauge system with a gauge of only 6.5 mm (a shade over a quarter of an inch), but in this country the most popular size is the 00-gauge system with a gauge of 16.5 mm. Table-top railways have a large following and achieve remarkable standards of detail and realism, but even with the addition of sound and smoke systems, the steam locomotive enthusiast may find the need for something more evocative of the real thing.

Between the two extremes there are various options for the locomotive fan to pursue, but what could be better than to ride behind a fully-working miniature replica of the real thing? Better still, to have built the model yourself from scratch. This way you can combine the dual pleasures of model-building and locomotive-running. On the track you can evoke all the sights, sounds and smells of live steam, whilst in the workshop you can gain immense pleasure from the construction of a working model which will be a lasting monument to your skill and patience.

And so we arrive at the subject of this book, the working model steam locomotive, examined from the view-point of the complete beginner. I do not intend to deal with any particular aspect in detail, to delve into the mathematical intricacies of valve-timing or the finer points of precision lathe work, but to give an overall view of the subject and some guidance as to how to get started in what can be a tremendously rewarding hobby.

A whole host of questions will occur to the beginner who has just visited the Wembley Model Engineering Exhibition or perhaps seen a couple of locomotives at work on the local model engineering club track. How do they work? Can I buy one complete, do they come as kits or do I have to build it myself? If so, what special tools and skills are needed and what does it all cost?

I am going to try and answer all these questions plus a few more, with the idea of easing the way for the budding model locomotive enthusiast and avoiding some expensive or embarrassing mistakes. I well remember an experienced toolmaker who proudly produced the assembled frames of his first locomotive at a local engineering exhibition for all to see. It was beautifully made and no one had the heart to tell him that he had riveted the hornblocks (the machined castings which carry the axle-boxes) to the wrong side of the frame plates. This kind of simple error can dishearten and turn away a person who would otherwise have turned out more than one fine model. Many locomotive models are started, but perhaps only one in five are ever completed. It is the aim of this book to reduce the casualties which fall by the wayside and become rusty collections of scrap in a forgotten corner.

Perhaps then, now is the time to sound a note of caution, and say that this hobby is not going to suit everybody. There are those with the means and the inclination to buy a complete locomotive; I shall have more to say about this later on, but the majority of people will think in terms of their own construction. Be warned. Do not attempt this kind of project lightly. The average person will spend 1,000 hours

LBSC's design 'Tich' is a popular choice with beginners. This very nice example is the small-boilered version, fitted with slip eccentric valve-gear.

Above *This is what model steam locomotives are all about. Here is a 3½ in-gauge saddle tank hauling its builder on a raised track in wooded surroundings. The scale mineral wagon is a nice touch since scale rolling stock is not often seen in these gauges.*

Below *This view of the cab shows all the controls on the boiler backhead. Note the miniature pressure gauge and neat pipework.*

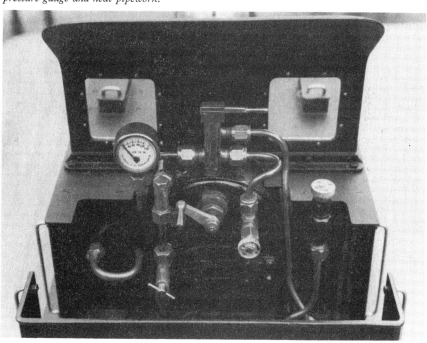

building the average locomotive. This is equivalent to six months' full-time work spent, for the most part, alone, closeted in the oily confines of a workshop.

Three factors affect the rate at which the model may be completed: time, skill and money. These are more or less infinitely interchangeable, that is, a skilled man will work faster than a beginner, a rich man will buy machine-tools to work quickly or buy completed parts such as the boiler. Lacking in either skill or money, then more time will have to be invested, but whatever your position within the triangle, there is only one ingredient for which there is no substitute... enthusiasm.

If, having read this book and perhaps some of those suggested in the bibliography, and talked to some experienced locomotive builders, you feel ready to try your hand, do not be daunted. Progress may at times be slow but rich pleasure can be gained from the completion of each part and from the solution of each problem as it arises, using the equipment at your disposal. I started out with a kit-built miniature lathe in the coal-shed of my student digs and eventually turned engineering into my career.

This shot shows the social side of the hobby at a typical model engineering society track. Various locomotives are standing on the steaming bays, waiting for their turn of passenger hauling as a train steams past on the main line.

Above *For those with a large garden, what could be nicer than a scene like this outside the back door! This 7¼ in-gauge railway has both steam and petrol traction. See Chapter 3 for some more pictures of the loco in the foreground.*

Below *Not recommended for beginners, but this beautiful 5 in-gauge 'Britannia' shows what can be achieved by the model locomotive builder. Only the angle iron rail head gives away the fact that it is a model, not the real thing. The detail work and finish are totally convincing.*

Left *Here is another 'Tich', this time the large-boilered version, straight from the assembly shop in bare metal. This example was built by a complete novice and whilst it would hardly win prizes at exhibitions, it is none the less quite nicely made.*

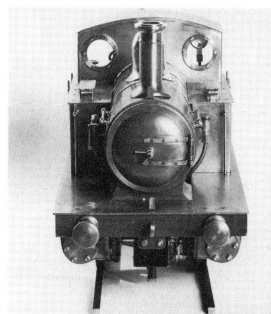

Bottom left *This view shows some nice platework on the engine. The blackened smokebox interior indicates that she has been trial steamed, prior to painting.*

Top right *Prominent in this front view of 'Tich' is the lubricator tank, just behind the buffer beam.*

Right *Cab view showing the boiler back-head. Note the pin coupling on the buffer beam, non-scale but a good plan for passenger hauling.*

A good deal bigger than 'Tich' is this 5 in-gauge 'B1' class loco, looking resplendent on the steaming bays. Large engines are impressive but remember that you will need larger machine tools to tackle their construction and a fair bit of strength to get them into your car boot!

A bit of history

Model locomotives have been around ever since the real thing but they became practical working propositions around about 1930 largely through the efforts of one man, L. Lawrence, who was born in 1882 and died in 1967.

'Curley' as he was known to his friends, wrote under the soubriquet of LBSC (taken from the London, Brighton and South Coast Railway) in the magazine *Model Engineer*, itself an institution dating from 1898. Lawrence worked for a time on the LBSCR but soon gave this up to devote himself entirely to the design and construction of 'little' engines as he called them. He despised the word model since, as he said, his engines were mere small editions of the real thing. Never an academic, always totally practical, LBSC wrote from the hard experience of locomotive construction, not from some theoretical standpoint, and although dated, his works are to be thoroughly recommended to the beginner. Perhaps most pertinent here is the collection of articles which appeared in *Model Engineer* during 1951 and 1952, and which were reprinted as a book called *Simple Model Locomotive Construction — Introducing Tich*. The little tank engine described in close detail in this work is a favourite beginner's engine, largely because of the very full description given in the book. LBSC established himself as the master of model steam locomotives with the so-called 'Battle of the Boilers' which began in 1922 and was more-or-less concluded at the 1924 Model Engineer Exhibition.

Above *This 5 in-gauge Canadian National 2-8-2 is running light here, but it can handle a one ton train without much struggle.*

Below *Steam locos come in all sorts of shapes and sizes. Some of the early American types give scope for an interesting model. This 2-4-4 tank is for 3½ in-gauge.*

The protagonists were on the one hand the commercial model locomotive makers, who favoured spirit-fired water-tube boilers and on the other, the lone LBSC who argued that a proper loco-type boiler with solid fuel was vastly superior. The battle raged initially in the columns of the model magazines at a time when model engines were expected at best to pull a dozen or so scale wagons around a garden railway. To suggest that a $2\frac{1}{2}$ in-gauge locomotive, fired by coal, could haul its live driver, was to invite the wrath of a somewhat entrenched commercial model locomotive industry.

Eventually a challenge was arranged to take place at the 1924 exhibition. The industry was represented by Bassett-Lowke, a famous name in the model world, who fielded a specially built three cylinder 2-8-2 named *Challenger*. In reply Lawrence produced *Maggie*, short for magnum opus, a four cylinder 4-6-2. This locomotive, however, exceeded the British loading gauge and was consequently disqualified, so that LBSC had to run *Ayesha* a diminutive 4-4-0, weighing 18 lb against *Challenger*'s 40 lb. The requirement was that each engine should haul a live passenger continuously for 15 minutes and this *Ayesha* did, taking 12 stone Lawrence on $22\frac{1}{2}$ laps of the track. *Challenger* however, hauling 9 stone, made 23 laps and Bassett-Lowke claimed a victory. A Pyrrhic victory indeed and proof, if proof were needed, that a tiny locomotive boiler can provide the steam to haul live passengers in the manner of the full-size article.

This story is well documented in the biography of LBSC by Brian Hollingsworth entitled *LBSC, His Life and Locomotives*. This book is to be thoroughly recommended to newcomers interested in the colourful past of model steam locomotives. It gives a fascinating insight into a talented but enigmatic figure, together with an account of his many designs both popular and rare.

More recently a number of people have produced successful locomotive designs but the most prominent is Martin Evans, whose works also appear regularly in the *Model Engineer*, the first, and still the principal, magazine for model engineering enthusiasts. Though perhaps more academic than LBSC, Martin Evans has produced a series of highly popular designs from 'Rob-Roy' a little $3\frac{1}{2}$ in-gauge 0-6-0 tank for beginners to 'Highlander' a $7\frac{1}{4}$ in-gauge version of Stainer's famous 'Black Five' 4-6-0 locomotive.

How do they work?

Here is a problem that faces the newcomer to model steam locomotives. Everyone knows how a steam locomotive works, don't they? How can anyone be so ignorant as to ask questions about a subject that any Englishman should know by instinct? Very easily, is the answer, so let us take a look at the insides of an engine. Model or full size the principles are much the same, perhaps the biggest difference being the relative size of the boiler tubes.

Let us start by looking at the cylinders and valve-gear. The basic principle is simple enough, steam under pressure is admitted to each end of a cylinder alternately, thus driving a piston to and fro. The longitudinal motion of the piston is converted to the rotary motion of the wheels by means of the piston rod, crosshead and connecting rod. Clearly, at each end of the stroke, no thrust is applied to the system so some means must be found to keep the loaded engine going. In single cylinder traction engines this job was done by the flywheel, the stored energy of which carried the piston over its dead centres. In locomotives, however, a second cylinder is provided and arranged so that it is working hardest, ie, is at mid-stroke, whilst the first cylinder is on one of its dead centres. The crank-pins of the two cylinders are therefore at 90° to one another and British tradition dictates that the right-hand crank should lead.

The majority of full size locomotives had only two cylinders, placed either inside or outside the frames, but quite a number had more. The famous 'Pacifics' (4-6-2 wheel arrangement) designed by Sir Nigel Gresley had three cylinders, two placed conventionally outside the frames and one inside, driving a crank-axle. These included the *Flying Scotsman* and the world's fastest steam locomotive *Mallard*. Many of the passenger engines of the Great Western Railway had four cylinders, two outside and two inside, making for interesting, if rather complex, models!

How is the steam admitted to the cylinders at the correct time and what happens when it is spent? This is sorted out by the valve, itself driven by one of the various types of valve-gear which I shall describe briefly in the chapter on locomotive chassis. Let's not worry about valve-gear now, just concentrate on the valves. These may be one of three types, slide, piston or poppet. The last of these, with valves rather like modern car-engine valves, is rare in full size, is extremely rare in miniature, and is not for the beginner unless a time-served watch-maker. Piston valves were almost universally used in full size practice and are quite common in models. They are desirable since they take only a small force to move them, but they are mechanically complicated and require close-tolerance machining to work well. Slide valves are often used in models, being simple and effective, and can be disguised to appear like the piston valves of the original. Have a good look at the Figure 1-2.

The steam chest is supplied with steam from the boiler via the driver's regulator. With the valve in the position shown, the front port is open, steam can pass to the front of the cylinder and the piston is being driven back down its bore. The back port, however, is open to the exhaust port via the cavity in the slide-valve, sometimes called a D-valve. Thus spent steam from the previous stroke is allowed to escape to the atmosphere. It does another important job on the way, but more of that in a moment, first we must mention the oft-discussed topics of lap and lead.

Lap is the extra length added on to the slide valve beyond the width of the inlet ports. The effect of this is to close the working port to live steam before the piston has reached the end of its stroke. As a rough guide, the valve is arranged to close the port after the piston has travelled 75 per cent of its course. This is termed a cut-off of 75 per cent. The reasons behind this apparently premature closure are varied, firstly there is an economy of steam gained by 'expansive working', as a puff of steam is admitted to the cylinder and allowed to expand against the piston. Secondly smoother running results, particularly at high speeds. This particularly helps model locomotives which tend to run at very high scale speeds.

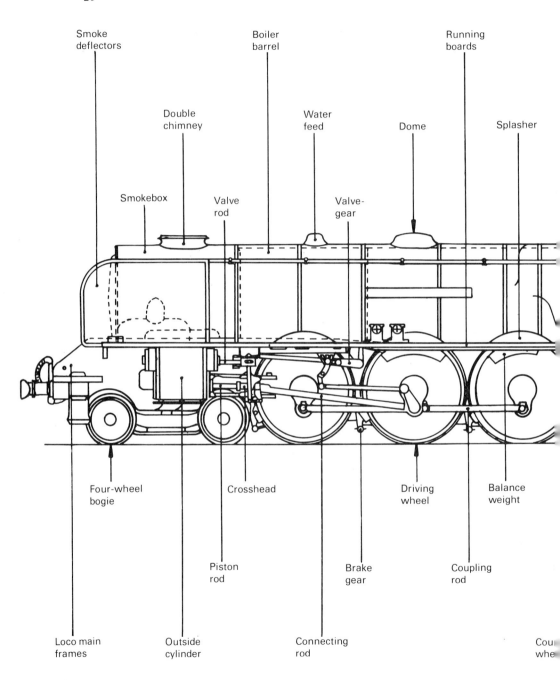

Smoke
deflectors

Boiler
barrel

Running
boards

Double
chimney

Water
feed

Dome

Splasher

Smokebox

Valve
rod

Valve-
gear

Four-wheel
bogie

Crosshead

Driving
wheel

Balance
weight

Piston
rod

Brake
gear

Coupling
rod

Loco main
frames

Outside
cylinder

Connecting
rod

Cou
whe

Figure 1.1: Drawing of a 'Pacific' type locomotive to show the major components. If you know nothing about steam locos, then study this carefully before proceeding!

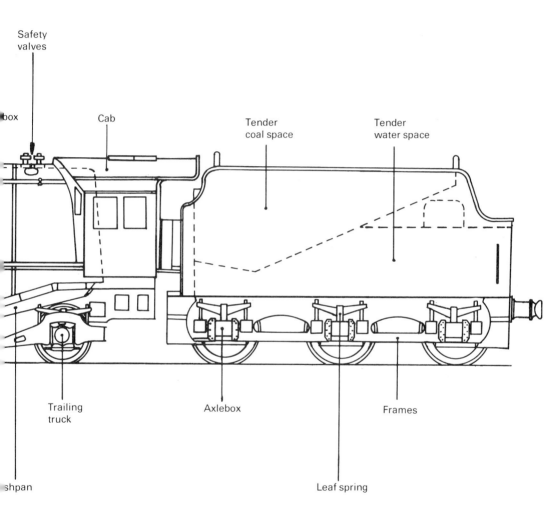

Safety
valves

box

Cab

Tender
coal space

Tender
water space

Trailing
truck

Axlebox

Frames

shpan

Leaf spring

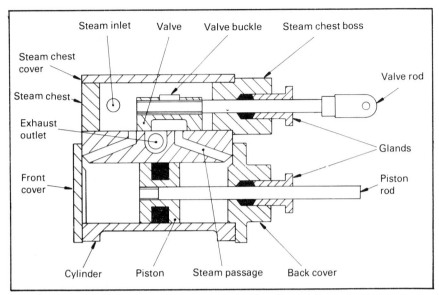

Steam inlet   Valve   Valve buckle   Steam chest boss

Steam chest cover

Steam chest

Exhaust outlet

Front cover

Valve rod

Glands

Piston rod

Cylinder   Piston   Steam passage   Back cover

Above *Figure 1.2: Side view section of a typical slide valve model cylinder block and steam chest. As drawn, the front of the cylinder is open to live steam whilst the back end is open to exhaust. Thus the piston is being driven towards the back of the locomotive.*

Below *Bernoulli's principle at work; the ping pong ball is suspended by the column of air from the vacuum cleaner hose which Suzie is holding.*

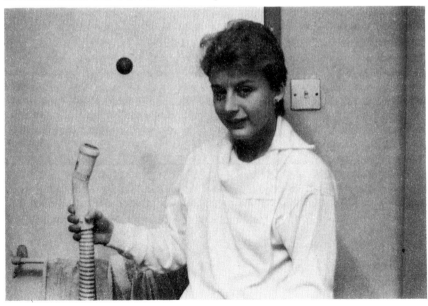

Figure 1.3: End view section of a slide valve cylinder and steam chest. This shows the cavity in the slide valve and how it connects through to the exhaust pipework.

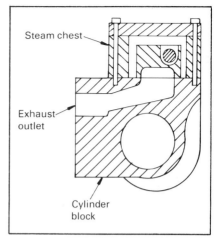

This cut-off value is not generally fixed, but variable by means of the cab reverser. This device, working through the valve gear, can be used to progressively shorten the cut-off from 75 per cent down to sixty per cent, forty per cent etc, until mid-gear is reached, a sort of neutral point. What happens thereafter? Continuing to bring back the reverser beyond mid-gear puts the system into reverse with the same variable cut-off available in mirror image. Cab reversers tend to be of the pole-type on older loco designs and the screw-type on more modern ones.

Clever stuff, isn't it? Thus the normal procedure, when starting a train, is to put the reverser into full forward gear, and then gradually open the regulator. In this condition maximum torque will be applied to the wheels for starting, but much steam will be used. As the train gathers speed the reverser should be brought back progressively to a running setting of around thirty to forty per cent, giving good steam economy combined with smooth running.

What about lead? This is rather like ignition advance in a petrol engine, it is the amount by which the steam is admitted to the cylinder ahead of dead centre, so that maximum pressure will be developed in the cylinder when the piston is in a suitable position to take advantage of it, and not way down near the end of its stroke. This is probably not so important in small locomotives with relatively short distances between steam chest and cylinder bore.

Now what about that exhaust steam with work still to do? This is conveyed to a nozzle placed in the smokebox, at the front of the boiler, and accurately set under the engine's funnel. The effect of the exhaust steam blowing up the funnel is to generate a vacuum in the smokebox, which draws air through the boiler, hence providing a draught to the fire. The harder the engine is worked, the stronger the exhaust blast, and the more the fire is drawn, generating more steam. This explains the sparks ejected from the funnel of engines starting out or on steep upgrades. The exhaust can lift the fire from its grate, draw it through the boiler tubes, and blast hot cinders from the funnel. This caused lineside fires in full size, and still does so with models! This extremely effective arrangement was devised by the Stephensons in the nineteenth century and has been a basic principle of locomotive design ever since. The physical phenomenon which causes the vacuum is known as Bernoulli's Principle and can be demonstrated with a vacuum cleaner and a ping-pong ball. Set the cleaner to blow and hold the nozzle vertically upwards. The ball can then apparently be 'balanced' on the jet of air.

Figure 1.4: Typical pole-type reverser as fitted to many models. The reach rod goes towards the front of the loco and connects with the weigh shaft to operate the valve-gear.

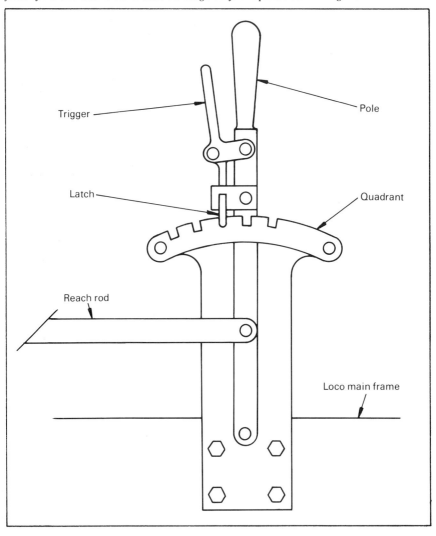

The reason is that the emergent air is at a lower pressure than the surrounding air and so the ball tends to 'fall' into this region. In the smokebox of the locomotive, the flue gases 'fall' into the exhaust steam column and are carried up and away from the engine by it. Incidentally we shall meet Bernoulli's Principle again in talking about injectors and it is the same phenomenon that draws fuel into carburettors and aeroplanes into the air.

The boiler

Model locomotive boilers follow full size design fairly closely and hence there are a number of designs available to the model builder, but they generally have the same main features, as shown in the diagrams. The fire is housed in a box surrounded by water, with a series of flue-tubes extending forward through the water-filled barrel to emerge in the smokebox. The fire rests on a removable grate, usually equipped with an ash-pan. Thus a large surface area is available to heat and boil water to provide steam. Most of the boiling is actually done in the region of the firebox, the boiler tubes serving more to pre-heat the water fed into the boiler.

Overleaf *Figure 1.5.*
Below *Figure 1.6: End view section of a basic boiler. This section is taken through the firebox but note that, for the sake of clarity, the firebox side stays have been omitted.*

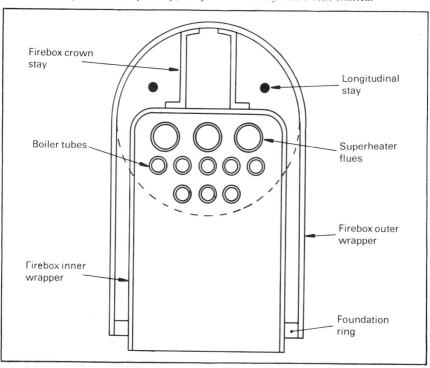

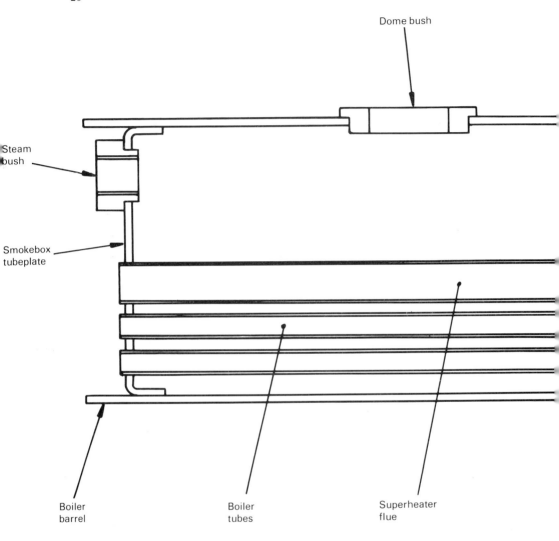

Dome bush

Steam
bush

Smokebox
tubeplate

Boiler
barrel

Boiler
tubes

Superheater
flue

Figure 1.5: Side view section of a basic locomotive boiler showing all the principal components. I suggest you refer to this diagram when reading Chapter 8.

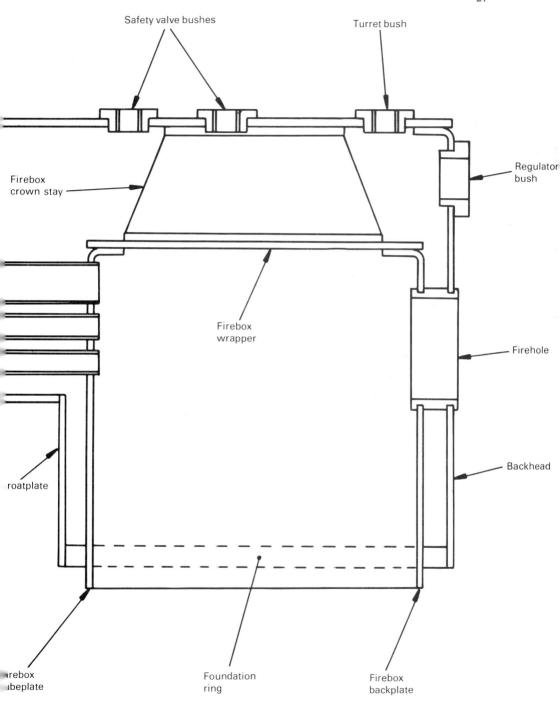

Safety valve bushes

Turret bush

Firebox
crown stay

Regulator
bush

Firebox
wrapper

Firehole

roatplate

Backhead

irebox
ubeplate

Foundation
ring

Firebox
backplate

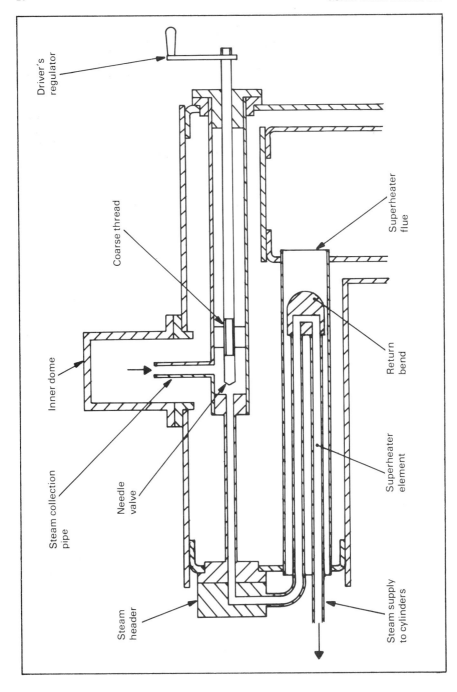

Above *This 'Tich' boiler has been sectioned cleverly to show its construction. Note how the firebox is surrounded by water space and how the fire tubes extend from it, through the barrel. Note also the hollow blower stay and the tiny stroudly-type regulator in the dome. This little kettle will produce enough steam to haul a live passenger continuously, given good quality coal and a deft hand on the regulator.*

Left *Figure 1.7: Side view section of basic loco boiler showing how steam is collected from under the dome, controlled by the driver's regulator from the backhead, and further heated by the superheater before being fed to the cylinders.*

This raises the next point. Clearly the boiler loses water as it is converted to steam and used by the cylinders and this must somehow be replaced. Furthermore, to push water into the working boiler, the feed pressure must exceed the boiler pressure. Perhaps surprisingly, this does not present much of a problem, and a beginner can happily make a simple pump with two ball-valves to feed water to a boiler. These pumps are generally placed between the locomotive frames and are driven by an eccentric, though some are driven directly by the crosshead of one cylinder. Alternatively, injectors or steam-driven pumps may be provided, but more of this later. Water for boiler feed may be stored in tanks of various types on the engine, hence the term tank engine. On larger engines, water and coal are carried in a separate tender.

The amount of steam fed to the cylinders is governed by the driver's regulator, controlled from the cab but often situated inside, or just below, the dome, where dry steam can be collected above the boiling water. This steam is usually further heated by passage through tubes placed in the path of the hot flue gases. This

A fine example of the boiler maker's art, this massive but relatively simple boiler is destined for a 3½ in-gauge narrow gauge locomotive. Note the many soldered stays which brace the flat sides of the firebox against internal pressure.

effect is called superheating. It does not greatly increase the power of the engine, since no increase in pressure results, but it does greatly improve water consumption, since more heat can be transferred by a given volume of superheated steam.

Lubrication is required by the cylinders and valves of a steam engine in the same way as any other type of heat engine. In models, this is usually achieved by means of a small oscillating cylinder pump driven by a ratchet mechanism from some part of the valve-gear. Thus every twenty or thirty turns of the wheels give one turn to the oil pump, pushing a globule of oil into the main steam pipe. The flow of hot steam blows this oil through to the valves and then the cylinders and eventually out of the funnel to splatter the face of the driver! Thick, syrupy steam oil is used and this 'melts' on meeting superheated steam.

Running gear

Thus far we have the main features of a steam engine, and now we have to think about the running gear; that is the wheels, and the multiplicity of ways in which these can be arranged. Early locomotives had a single driven axle and axles with undriven smaller wheels front or back, but as locomotives grew larger and heavier it became desirable to spread the drive to four wheels linked by coupling rods, and later on to six and more coupled wheels. At the same time speeds increased

The front end of this boiler shows 24 fire tubes, and two superheater flues. The large hole at the top is for the steam header, the smaller one below and to the left of it is for the blower steam supply.

dramatically and fixed wheelbases proved rather harsh on both engine and track. The solution to this was to provide hinged two-wheel pony trucks or four-wheel bogies ahead of the driving wheels, centralized by large springs, so that the force guiding the locomotive into a curve was spread between the leading bogie or pony truck and the driving wheels. This makes for a flexible engine which will ride smoothly over the irregularities of the track, and is equally applicable to models.

Each axlebox on the locomotive is also sprung vertically both to soften the ride and keep each wheel loaded on to the rail head. This leads to a requirement for a certain amount of play in the running gear to allow for the irregular vertical movements of the wheels.

Brake gear on a model locomotive is largely cosmetic because the train heavily outweighs it, however, scale manual or vacuum steam brakes can be fitted as desired along with all the other bits and pieces like hand rails, lamp iron, name and number plates, that form part of the essential character of a steam railway locomotive.

When it comes to the running of a model steam locomotive, things work out much like full size except for initial steam-raising. To begin with suitable pieces of wood soaked with paraffin are piled into the firebox and then a special blower is fitted to the locomotive funnel to pull air through the fire. This may be an electric fan arrangement or an ejector using compressed air. With the external blower running, a lighted match is tossed into the firebox, whereupon clouds of noxious white smoke emerge from the blower. If all is well, this ceases as the wood lights up. After a few minutes small pieces of coal are added on top of the wood, then steadily more and larger pieces. As steam pressure builds up to 20–25 lb on the pressure gauge, the external blower is removed and the locomotive's own steam blower opened up. This is just a fine jet of steam emerging beside the blast-pipe and directed up the funnel to draw the fire when the locomotive is stationary and there is no exhaust blast. This blower is controlled by a wheel valve in the cab. Incidentally, one used to hear in the old days of mainline steam of so-called 'blow-backs' occurring on engines, often resulting in fatal burns to the crew. The cause of this was simply the drivers' failure to open up the blower of a coasting engine, with the result that the slipstream could blow the flue gases back into the cab when the unfortunate fireman opened the fire-hole door to put in more coal. So, remember to open up the blower when you stop at a station or a red signal!

Chapter 2

What is around?

Working live steam models have been built experimentally in very small gauges and are available commercially in 00-gauge (16.5 mm), fired by tiny butane burners. At the heavy end of the scale, 15 in is the largest gauge of 'model' locomotives, anything beyond this being considered as narrow-gauged, full-size. The classic examples of 15 in-gauge railways are the Romney, Hythe and Dymchurch, and the Ravenglass and Eskdale. In both cases, rather over-scale locomotives are used to pull heavy commercial trains.

I shall define the scope of this book as covering those gauges capable of hauling live passengers, but which can be constructed by an amateur equipped with an average workshop (see Chapter 4). This, in practice, means just three gauges, $3\frac{1}{2}$ in, 5 in and $7\frac{1}{4}$ in. These three gauges derive from the scale equivalents of $\frac{3}{4}$ in:1 ft, 1 in:1 ft and $1\frac{1}{2}$ in:1 ft respectively, with some slight liberties taken in order to 'round off' the figures. Thus 5 in-gauge scale locomotives are sometimes made to $1\frac{1}{16}$ in:1 ft, resulting in a slightly larger model. In the USA the gauge of $1\frac{1}{2}$ in scale is also commonly set at $7\frac{1}{2}$ in.

Gauge	Loading gauge to British Standard 9 ft × 13 ft 6 in	Average loco weight	Average train weight	Scale in inches to the foot
$2\frac{1}{2}$ in	$4\frac{3}{4}$ in × $7\frac{1}{8}$ in	50 lb	200 lb	$\frac{17}{32}$ in
$3\frac{1}{2}$ in	$6\frac{3}{4}$ in × $10\frac{1}{8}$ in	100 lb	1,000 lb	$\frac{3}{4}$ in
5 in	$9\frac{1}{2}$ in × $14\frac{1}{4}$ in	250 lb	1,750 lb	$1\frac{1}{16}$ in
$7\frac{1}{4}$ in	$13\frac{1}{2}$ in × $20\frac{1}{4}$ in	500 lb	5,000 lb	$1\frac{1}{2}$ in

From the diagram you can see that scale models built to continental or American prototypes will be noticeably larger than those built to British outline, which is something to bear in mind when choosing a prototype to model. Some years ago $2\frac{1}{2}$ in gauge ($\frac{1}{2}$ in:1 ft) was very popular, being cheap in terms of materials and easier to transport for those without cars. Certainly $2\frac{1}{2}$ in-gauge engines will haul live passengers, indeed this was LBSC's favourite gauge and today is undergoing something of a revival under the auspices of the $2\frac{1}{2}$ in-gauge society. However, I feel that something a little larger is a better proposition for the beginner. Fits and tolerances are just that bit more critical in the smaller gauge and the parts that much more fiddly to make. When complete, the engine is going to be a little marginal on power, added to which there are not many $2\frac{1}{2}$ in-gauge tracks in this country.

Incidentally, one should not decry the power of the small gauges entirely, since LBSC produced an 0-gauge locomotive ($1\frac{1}{4}$ in gauge) of British standard gauge outline, named *Sir Morris de Cowley*, which would pull a live passenger. By the same token perhaps the beginner should not shun the larger gauges. A good friend of mine started out as a complete novice and designed his own $7\frac{1}{4}$ in-gauge tank engine, cast his own cylinders, wheels, dome, etc and made every component apart from the piston rings and pressure gauge. No mean feat for a man with

Figure 2.1: This shows the tremendous difference in the size of loco, full size or model, built to either British or American loading gauge.

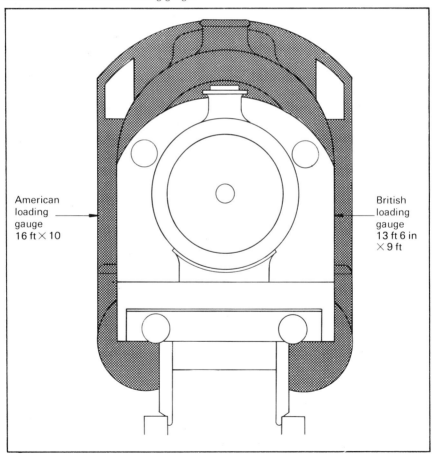

American
loading
gauge
16 ft × 10

British
loading
gauge
13 ft 6 in
× 9 ft

family commitments. (See photographs in Chapter 3.)

However, when all is said and done, the best bet for a beginner is probably 3½ in gauge. A tender engine of this gauge will handle a load of four or five passengers quite comfortably on the track, but will be light enough for one man to lift out of the car boot and place on the rails without slipping a disc. At the same time the components will be within the design capacity of the ubiquitous 3½ in centre-height lathe which is a significant point for those planning to build themselves. It seems to me that model engineers spend nine-tenths of their time in devising ways to press their machines to handle jobs way beyond their nominal capacity. Fine if you enjoy tackling that kind of problem, but for me, life is too short for this kind of carry-on. If you must have a big engine, either set up with large equipment or arrange to borrow machines by some means, fair or foul. There will be more on this topic in Chapter 4.

So, what designs of locomotive are available to us in model form? The answer for those who can commission the locomotive of their choice will be that virtually any design can be produced. Buy a book illustrating the types of locomotive used in this country, or anywhere in the world for that matter, and make your choice. The variety of shape, size and colour is enormous, from the very early single wheel locomotives of the 1830s, such as Stephenson's *Rocket*, through to the huge articulated American engines of the 1940s, just prior to the diesel invasion. I imagine though, that if you are prepared to spend the sort of money required here, then you will have a pet locomotive close to your heart for one reason or another, and the choice will be an easy matter. By and large models in Britain are of scale, or at least near-scale, outline based on standard-gauge British full-size engines, however, there are plenty of exceptions. Foreign locomotives are frequently modelled, American prototypes being quite popular whilst there is an increasing interest in narrow-gauge designs. These give relatively large and consequently powerful engines, but of simple design and construction, usually as 0-4-0 tank engines.

Free-lance loco designs are also quite popular and leave scope to choose whatever outline and detail appeals to the builder. A favourite subject of the late LBSC was the 'might-have-been' locomotive, something to challenge the mighty diesels in power and speed. Indeed, a lot of his ideas ran in miniature form but not full size.

Now is perhaps a good time to say something about wheel arrangements and general designs of locomotive, for the benefit of complete novices. Broadly speaking there are five classes, express, mixed traffic, goods or mineral, tank and articulated. This latter category hardly featured in British full-size practice, but was, and in some cases still is, used on many railway systems abroad. In general it is the arrangement of the wheels which decides the category into which a locomotive falls, and this arrangement is described by giving the number of leading, undriven, wheels, the driving wheels and then the trailing, undriven, wheels. Thus the famous *Flying Scotsman* had a 4-6-2 arrangement, that is four undriven bogie wheels, six coupled driving wheels and two trailing wheels under the firebox. Using the same designation, Stephenson's *Rocket* was an 0-2-2,

A fairly typical 0-4-0 narrow gauge engine. This shows how a large but simple model results from using a narrow gauge format. This particular model is unusual in that it has Walschaert's valve gear instead of the Hackworth gear common on this kind of loco. (Patrick Advertising Associates.)

having no leading undriven wheels, a pair of drivers, and two trailing wheels. *Evening Star*, the last steam locomotive built for British Railways, was by contrast a 2–10–0, whilst the huge articulated Union Pacific 'Big Boys' in the states, with their two sets of driving wheels, were designated by 4–8–8–4.

Since British prototypes are the most popular choice in this country, I shall confine this brief description mainly to British practice.

Express engines

Early express engines had only a single pair of driving wheels or just four coupled wheels, often with another non-driven axle fitted directly to the frames, either in front of, or behind the drivers. Examples of this are the famous LNWR 2–4–0 *Hardwicke* now preserved in the National Railway Museum, York and *Lion*, the Liverpool and Manchester Railway 0–4–2 immortalized in the film *Titfield Thunderbolt*. Both of these exist as designs in model form with drawings and castings and so forth readily available, both incidentally by the old master LBSC. Similarly available examples of single-wheelers are the 'Stirling 8 ft Single' of the GNR and the so-called 'Johnson Spinners' of the Midland Railway, both with 4–2–2 wheel arrangements. Like their full-size counter parts, single-wheelers can tend to suffer from lack of adhesion but careful attention to springing can minimize this problem, indeed a 5 in-gauge 'Stirling 8 ft Single' has been the victor at IMLEC in recent years (see Chapter 9). Sanding gear extended the life of full-size single-wheelers but this is not recommended for models since it inevitably gets into the works, where it does no good at all!

These old engines have classically elegant lines and make a fine sight loping

This 5 in-gauge 'Britannia' 'Pacific' is an excellent example of an express-type engine.

along with their huge driving wheels revolving at a leisurely pace. Later on, development of the four-wheel leading bogie made the 4-4-0 arrangement practically standard for express engines, examples being the 'Crimson Ramblers' of the Midland Railway and the 'Schools' class of the Southern whilst *City of Truro* of the GWR was the first steam engine to exceed 100 mph. This arrangement also embraces the classic American locomotives seen in all the best 'Westerns', with their huge spark-arrester funnels, cowcatchers, lamps and bells, making for a colourful and quite popular model choice.

The next development was the addition of a trailing set of wheels under the firebox to make a 4-4-2, the so-called 'Atlantic' wheel arrangement, after the first locomotive to be built thus. A pretty example of this is the GNR Atlantic, usually finished in the attractive apple green of the LNER in the 1920s and '30s, again available as an LBSC design. *Ayesha*, LBSC's locomotive in the 'Battle of the Boilers' was of course an Atlantic.

Next came the 4-6-0, perhaps the most common of wheel arrangements both for express and mixed traffic work. The GWR used this for most of its passenger engines, the 'Saints', 'Stars' and 'Castles', and even for its largest express locomotives, the famous four-cylindered 'Kings'. On the LMS, the 'Royal Scots' were 4-6-0, on the Southern there were the 'Lord Nelsons' and 'King Arthurs' and on the LNER the 'B1s'. Once again, many of these are on offer as tried and tested models in various gauges.

Finally we come to the classic British express locomotive, the 4-6-2 'Pacific', used until the end of steam in 1965, to haul long-distance fast passenger trains. Sir Nigel Gresley favoured the use of three cylinders for his most famous works, the 'Flying Scotsman' 'A3s' and the streamlined 'A4s' such as *Mallard*, which still holds the world speed record for steam traction. The 'A3s' make very popular models, the 'A4s' less so because of the problem in making an accurate replica of the steamlined casing. The LMS ran my own personal favourite locomotives, the 'Duchesses', four-cylindered 'Pacifics' designed by Sir William Stanier. In terms of tractive effort, these were the largest non-articulated engines to run in Britain, making for a powerful, if rather complex model. The British Railways 'Britannia' design is a widely modelled 'Pacific', whilst an interesting project would be a miniature version of the curious Bulleid 'Pacifics' of the Southern Railway with their oil-bath enclosed, chain-driven valve-gear.

Mixed traffic engines

This group, as the name suggests, were maids-of-all work, running both freight and passenger trains and perhaps forming the backbone of traction on the railways. The definitive example here is the extremely successful Stanier design known as the 'Black Five', the most numerous of all British steam locomotives, built first of all for the LMS. The Great Western equivalents were the 'Halls' and 'Manors', whilst the LNER produced the Thompson designed 'B1' Class. The 2-6-0, or 'Mogul' arrangement was used for mixed traffic as was the 2-6-2, though neither proved very popular in this country.

Freight engines

In order to pull a heavy load at a slower speed than the express locomotives, this group tended to have a larger number of smaller wheels with diameters around 4 ft rather than the 6 ft of the fast locomotives. In the early days 0-6-0s did most of the freight handling and though very numerous with types such as the 'Jones Goods', 'Dean Goods', 'Derby' Class '4F' and the strange Southern Railway 'Q6', these unglamorous engines are not often seen working in miniature form.

A little later came the 0-8-0 and then the 2-8-0 or 'Consolidation' freight engines such as the Stanier '8F' of the LMS or the Class 4700 of the GWR. One problem from the model point of view is the length of the wheelbase of these types of locomotive which means that they can only negotiate curves of fairly generous radius. This is even more of a problem for ten-coupled engines such as the *Evening Star*, the British Railways '9F' class, last steam locomotive built in this country for the standard gauge. This type is nevertheless quite popular and can be found running in both $3\frac{1}{2}$ in and 5 in gauge.

Tank engines

This group is so-named because, instead of pulling a separate tender with a supply of water and coal, they have one or more tanks and coal bunkers mounted in various ways on the locomotive itself. This made for a compact and fairly cheap engine with plenty of adhesive weight on the driving wheels. The shape and position of the water tanks defined the various types of tank engine. Those with boxes either side of the boiler resting on the running boards were side-tanks, those with a curved tank resting over the top of the boiler were saddle-tanks. Many Great Western 0-6-0s had tanks mounted on the sides of the boiler but not reaching down to the running boards so that there was access to the inside valve-gear. These were pannier-tanks. Some of the earlier engines carried water between the main frames and these were called well-tanks. A relatively small

Above *Martin Evans designed this 2-8-0 freight loco for 5 in-gauge and named it* Nigel Gresley *in honour of the designer of the full size engine.*

Below *The British Railways 'Class 2' was a classic mixed traffic loco. This fine model in 5 in-gauge is to a design by Don Young.* (David Palmer.)

quantity of coal was placed in a bunker, usually behind the cab, thus with a relatively small supply of water, tank engines were generally used for short-haul duties, shunting, suburban trains, banking and so forth. There was a huge variety of these locomotives with virtually all the wheel arrangements of the tender engines, plus the diminutive 0-4-0 which was rarely seen with a tender.

Tank engines make popular models for several reasons. Firstly the extra work in making a tender is avoided, though of course there is considerably more platework to do on the locomotive. Secondly, they are comfortable to drive since one does not have to lean forward across a tender to reach the cab controls. The extra adhesive weight of tanks and water is useful when pulling heavy loads. On the minus side, tank engines tend to be rather inaccessible with all the works hidden away behind the platework, making for more difficult assembly and maintenance, besides which, having made all those intricate pieces, why hide your light under a bushel, or a water tank?

Beginners very often decide on a tank engine as being small and simple for a first engine, but this is in many ways a mistake. Apart from the tender chassis, there are nearly as many pieces to make in a tank locomotive and believe me, it can become pretty frustrating trying to assemble a $3\frac{1}{2}$ in-gauge tank locomotive, the more so with inside valve gear and worse still with both inside cylinders and gear as with LBSC's design 'Molly', the old LMS '2F'. Look at the design carefully and think whether your fingers are going to be able to fit that machinery together reasonably easily. Included under tank engines is also the 'Sentinel' type of geared locomotive which makes a fairly popular model. These had vertical boilers supplying steam to a two-cylinder engine unit which drove the wheels via chains. The Stuart-Turner stationary steam engines are often adapted for this purpose.

Articulated types

The purpose of the articulated locomotives was to provide a single unit of high

A pretty little 5 in-gauge tank engine. Compact and manageable, but quite a powerful puller.

An imposing example of an articulated loco. This is a 7¼ in-gauge edition of the enormous 2-8-8-2 'Mallet'-type locos built by the Union Pacific Railroad in the USA and aptly called Big Boys.

power, but capable of negotiating relatively sharp curves. They mostly fall into one of two categories, the 'Beyer-Garratts' and the 'Malletts'. The 'Garratt' consists of two complete locomotive chassis set out like the bogies of a coach, with a large boiler bridging across them on a rigid frame. Flexible ball-joints were provided in the steam and exhaust pipes to allow for the motion of the two locomotive chassis, whilst water was generally carried on the leading engine and coal or oil on the trailer. A powerful and flexible locomotive with excellent accessibility to mechanics and ashpan resulted, and the same is true of model versions, the disadvantages being that one is virtually building two locomotives at the same time. Gresley designed a six-cylinder 'Garratt' for LNER to haul heavy coal trains but perhaps the most notable use of the 'Garratts' has been in South and West Africa, on the metre-gauge lines.

The 'Mallett' type was very popular in the United States, resulting in some enormous engines capable of developing as much as 7,000 horsepower continuously, as compared with the Stanier 'Pacifics' which produced about 1,700 horsepower. The Malletts again had two locomotive chassis but one was fixed to the boiler assembly whilst the forward unit was free to move round to accommodate curves. These tended to be rarely modelled, probably because of the sheer enormity of the project. Other articulated types include the 'Shays', 'Heislers' and 'Climaxes' built to run on very rough and sharply curved tracks for logging or mining railways. These had their boilers slung between two bogies which were powered by a system of shafts and gears from an engine unit mounted on the boiler. These can be seen running in miniature form and certainly make a novel subject.

Thus the variety of steam locomotives is practically endless from the simple to the complex, from large to small, the choice of locomotive is a matter of continuing and heated debate for the modeller, but the decision as to which to build or buy lies with you, the enthusiast.

Chapter 3

Beg, build or buy

Buying new

There has long been a general feeling amongst the model locomotive fraternity that simply going out and buying a new locomotive is somehow 'not cricket', however the world is changing and this is an option which bears serious consideration. It is important for the new locomotive enthusiast to decide where his aims lie, whether it be in construction or just the pure running of a locomotive. If perhaps you just want to run an engine, enjoy yourself on the track and give rides to the kids, then that is fine. Taking into account the cost of workshop equipment (see Chapter 4) and materials, and the time taken in construction, your best solution may be to don a smart suit and simper to your bank manager. For those willing to spend a goodly sum, there are several professional locomotive builders in this country who will assemble the locomotive of your dreams in the colour of your choice but, perhaps not surprisingly, the cost is high. Even a professional will take more than 1,000 hours to finish a locomotive and he must cost himself out at more than £5 per hour and probably nearer £10 per hour to make a reasonable living. Thus a fully detailed, painted and lined 5 in-gauge passenger locomotive such as the ever popular *Flying Scotsman* or a Great Western 'King' Class will set you back £15,000. A moderately serious sum of money, however, it should be said that if the engine is carefully used and well maintained then it will not only maintain its value, but possibly increase. It will also last almost indefinitely. I have heard it said by a professional model locomotive builder that a large proportion of his work ended up in bank vaults, merely as a hedge against inflation for wealthy investors. Rather a waste of a man's devoted efforts, but an economic fact of life. Going to extremes, the world's largest steam locomotives were the so-called 'Big Boys' of the Union Pacific Railroad in the States. One of these was recently built in $7\frac{1}{4}$ in gauge for the Forest Park Railroad, Dobwalls, Cornwall, at a cost of £75,000. However, few beginners will really wish to run an engine weighing just under two tons!

One of the great advantages of buying from a known professional builder is that one is assured of the quality of the machine. Not only should the overall finish be excellent but the mechanics should be finely engineered with no 'tight spots' as the wheels revolve, the valve timing should be spot-on and the exhaust beats consequently even. Safety is a matter of prime concern and no one would wish to sit behind a couple of gallons of water at 150°C and 100 psi held in a pressure-vessel of dubious quality. Thus it is a pleasant reassurance that the professional will have supplied a well-designed and properly constructed boiler. No doubt the builder will also be happy to give the new owner plenty of instruction regarding the running and maintenance of the engine and should anything go wrong, then the builder will probably help out.

As a slightly cheaper alternative to commissioning a specific design, one can scan the columns of the model engineering press for adverts offering locomotives now under construction, ready for delivery in a few months' time. The unit cost will be much lower if a professional builds, say three engines simultaneously, so it makes sense if he puts together a batch, then offers these for sale. You may not get the exact design you wanted, but the price will be a little lower. Guarantees similar to commissioned engines will probably apply.

Next down the line of new purchase are the semi-professional or non-professional people who offer locomotives for sale 'steamed for trials only'. This group may embrace engines of very varied quality from good to poor, and so some degree of judgement is called for. This is going to present something of a problem for a beginner who would be well advised to seek advice. One course of action is to submit the locomotive, by agreement with the vendor, to one of the recognized and reputable locomotive dealers in this country for an impartial valuation. There may well be some sort of charge for this service, but as with the AA examination of used cars, a financial blunder could be avoided. Another possibility is to seek local advice by contacting, and preferably joining, the nearest model engineering society. Members of these clubs are often all too willing to voice their opinions, but, amongst all the talk, some sense will no doubt emerge and friends found who will help assess a locomotive. Guarantees here will also be variable, some vendors being more willing to sort out problems on troublesome engines than others. So as always, *caveat emptor*.

Mention of model dealers of course brings in another aspect. There are a number of organizations in this country working on a more or less professional basis dealing in models of all sorts, but locomotives tend to predominate. The larger of these produce a stock catalogue which is available on request, making interesting reading for the new potential locomotive buyer. These should give a good idea of what is available, and at what price, though the majority of engines offered will be second-hand (see next section). It would be prudent to enquire of the dealer what guarantees apply and what instruction might be available to the purchaser.

This aspect is not open to those who buy a locomotive through yet another route, that of the auction room. Several model auctions are held during the year and though second-hand locomotives again predominate, new locomotives do come up, and can be acquired at a bargain price in some instances. The snag is, of course, like the infamous car auctions, that one essentially buys as seen and the beautifully painted locomotive in the sale room may prove to have a leaking boiler-barrel seam on the track. Once again, seek advice, take someone with experience along with you and be sensible, auctions can do strange things to people!

The price one has to pay for a new locomotive from the dealers or semi-professional builders will vary with quality. It may be in the £5,000 to £10,000 bracket of the full professionals, or it may be much less. An acquaintance of mine recently bought a 5 in-gauge 0-6-0 tank locomotive brand new at auction for £1,000. This was not a fine scale job by any means, but a chunky, well assembled engine capable of hard work on the club open days. It was however a bargain purchase and £2,000 to £3,000 may be nearer the mark for a 5 in-gauge engine.

Buying second-hand

Having considered buying a brand new locomotive, I suspect the majority of people will be surprised, if not shocked, by the likely costs and so will promptly turn to the second-hand market.

There is quite a sizeable market in this country for second-hand model steam locomotives and for some reason this tends to be somewhat covert. The indications are that demand exceeds supply and therefore that deals are concluded without ever reaching the advertising stage, and prices are thus on the high side. Looking at the columns of models for sale in the magazines however, I find prices depressingly low in many instances and so am at a loss to explain the vaguely shady nature of this market. The economic recession has no doubt pushed models on to the market which is perhaps sad, since a locomotive which took the builder hundreds of painstaking hours to make, may fetch only a few hundred pounds and is lost forever. Another economic fact of life.

The first source that the beginner will most likely examine is the adverts in the model engineering magazines and perhaps *Exchange & Mart*. Reading through recent back issues of these magazines at the library or local model engineering club will give a fair indication of what is available and at what sort of price.

The next source to try is the dealers. These also advertise in the model press and

This 5 in-gauge 'K1' was bought second hand at an auction and after a little tuning work, has proved to be a very fine engine.

as mentioned under new locomotive purchase, some provide a catalogue of their wares which will also make interesting reading for the new locomotive enthusiast. As commercial suppliers these people should provide some form of guarantee and you should enquire how far this extends beyond the 'statutory rights'.

The next course of action and one worth pursuing purely for the entertainment, is to go to one of the model auctions as mentioned before. Once again, these will be advertised well before the event and this gives time to make arrangements, both financial and social, for the great day.

Finally, word-of-mouth can put buyer in touch with vendor on many occasions, so talk to people at model engineering clubs and visit other societies when possible to meet potential sellers. Naturally, if you can eliminate the middle man in the form of dealers or auctioneers, then a better bargain can be struck for both parties.

What should you look for in assessing a second-hand model steam locomotive? Like everything else, you will in general get what you pay for, and low price means low quality or long prior use. Your chief worry with a used locomotive should be the state of the boiler — whilst a silver-soldered copper boiler will last a long time, careless use and storage can result in dangerous corrosion. The supply of a boiler certificate is useful but not a cast-iron guarantee and the ideal solution is to ask your local model engineering society to witness a test on the engine. However, this is rarely practical. I think one should insist on a steam trial on the track at least. This will reveal any obvious fault in the boiler and mechanics and also give a chance to assess the seller as well! It may be that now is the time to say a little about boilers from the safety aspect, leaving discussion of the boilers in general to Chapter 8.

Model boilers work at between 60 and 120 psi, generally 80 psi, and contain perhaps a gallon of water, at between 134°C and 167°C, thus they must clearly be properly designed, constructed, tested and maintained. Modern designs have an ultimate safety factor of ten, so that a boiler stressed for 100 psi operation should ultimately fail at 1,000 psi. Construction now employs brazing or silver-soldering throughout, with the use of soft-solder confined to the caulking of the stay-heads only. New boilers are checked by means of a hydraulic test, that is, the boiler is suitably plugged at the various bushes for dome, water gauge, etc, and then completely filled with water. A simple hand pump is then connected and two or three strokes will bring the pressure up to twice working pressure. If the boiler should fail then there is no danger from the incompressible water. In this pressurized condition the new boiler can be checked for weeps or bulges and if all is well a certificate can be made out.

A steam test at one and a half times working pressure is often also carried out before commissioning a boiler. Subsequently the boiler should be given a hydraulic test every year or two years, to one and a half times working pressure. This can be done on the locomotive without removing any fittings other than the safety valves and pressure gauge. The authority to supervise this operation and issue boiler certificates will usually be the committee of the local model engineering society, another reason for the beginner to seek out and join the club. A boiler certificate is required for insurance purposes by most clubs, and will generally be recognized by other clubs, should one travel around with an engine.

Happily, because of the self policing carried out by the model engineering fraternity, accidents with boilers are very rare, in fact, virtually unknown, but should something serious occur, no doubt it would not be long before some fanatic legislator sought to terminate our enjoyment.

Essentially, general appearance will be the main indicator of the condition of an engine. If the overall impression is of cleanliness and neatness then things are promising, but if the machine is dull and dirty, be careful. Appearances can however be deceptive and I came across a 5 in-gauge 0-4-0 tank engine recently which appeared very shabby, being a dull oil-stained black from smokebox to dragbeam, but closer examination revealed sound workmanship and potential for overhaul to a high standard. More mention will be made of renovation later, in this section.

Now let's consider how to look at a used engine in some more detail, assuming that we like what we have seen so far. First of all, put it on a short section of track or a suitable flat surface, minus the tender if it has one. Hold it by the buffer or dragbeam and roll the locomotive gently to and fro, a couple of revolutions in each direction, and feel the resistance. This should be slight but positive due to the packing of the cylinders and glands, so that with a gentle push the engine will roll on for a revolution or so. It should also be quite even with no sticky or tight spots as the wheels revolve, which would indicate poor fit of the motion-work. There should also be some hissing noises from the smokebox as the pistons pump air to and fro and this will be more pronounced with the cylinder drain cocks closed. At the same time the mechanical parts should be silent, other than a probable click from the lubricator ratchet-drive. Beware of systematic knocks and look for the cause. Worn coupling and connecting rod bushes are relatively easy to replace but worn axleboxes present more problems as do worn pistons and cylinders. Check each axle for front-to-back play, particularly the driving axle and check the coupling and connecting rods for up-and-down play. All of these may be several 'thou' (thousandths of an inch) but not much more. This will be more tricky with an inside cylinder locomotive which is best laid on its side to get at the 'works'. Check each component of the valve-gear in the same way and then look at the lost motion in the valve rods with the reversing lever in full forward gear. Push the valve rod to and fro with a thumb and forefinger and assess the resulting play. About 15 thou or even 20 thou of play is reasonable but beyond this valve events will start to become erratic resulting in poor performance and increased wear of the system. Smaller gauge engines will clearly suffer more proportionately from a given amount of play.

Examine the crosshead and slide bars for wear and hope for less than 5 thou of play in the crossbar across the slide-bars. Make sure the piston rods are well secured to the crossheads because a loose fit here can mean trouble. While in this area, have a look at the piston and valve rod glands. See that they can be tightened down with little more than finger pressure to seal against their respective rods. Check the rods for heavy scoring through the gland as well. Now work the cab reverser through its full range and see that this gear works smoothly. Next find the lubricator; this is commonly found between the frames immediately behind the front buffer beam and covered by a portion of the running boards, but may lurk somewhere under the cab floor or alternatively may be perched boldly on the running boards beside the front of the boiler barrel. Whichever the case, remove its lid and see that the little oscillating pump inside works round as the wheels are revolved and that the ratchet drive does its stuff when the crankshaft of the pump is turned by pressure on the ratchet wheel with a finger tip. The oil tank should, of course, be clean and not full of ash, which does the valves and cylinders no good at all. If you fail to find the lubricator, then the engine probably has a hydrostatic oil feed, so look for the sight glass in the cab and check the system when the engine is in steam (see later). If the engine has one of the old type of simple hydrostatic feed lubricators fitted somewhere on the main steampipe, regard the whole deal with suspicion.

Finally, on the mechanical bits, look at the axle-driven feed pump which will likely be between the frames, somewhere about the middle of the engine, driven by an eccentric. Check for slop in the eccentric strap and the pump-arm. The main gland should tighten easily on its packing in the same way as the piston-rod glands. Obvious water leaks from the pump suggest problems. Crosshead-driven pumps are simpler, since no eccentric is involved but glance at the ram and gland.

We have mentioned the main aspects of the boiler but not its various fittings. The cab regulator should work smoothly from stop to stop with an even resistance, whilst the water gauge and its blowdown valve should be clean and fur-free. The various wheel valves should work freely and close firmly, and the firehole door should open easily and stay closed when required. Overall, the fittings should not show signs of leaks or heavy incrustations with chalky deposits.

Another 'overall' to look at is the plumbing work which should be neat and not battered and kinked. Again look for leaks and white deposits and see that the union nuts are in good condition, rather than marked by the use of wrong-sized spanners. The unions holding on the injector, if the locomotive has one, are prime candidates for a battering. Injectors, by the way, are lumpy brass things with four pipes leading to them, usually sited low down by the cab.

Going up to the front end, open the smokebox door and, with the aid of a torch, have a look around. The door itself should shut air-tight onto the smokebox ring, so examine the seating of these two items for lumps and bumps which would tend to destroy the vital smokebox vacuum. Actually you will not see very much inside the box, since everything will be plastered with soot and ash, but you can tell if the owner keeps this area clean, and you can look at the general condition of the superheater-headers, steam pipe, blast pipe and blower nozzle. Incidentally, if all these terms go over your head at this stage, read on, because explanations will appear in due course and at the end, all should be reasonably clear.

Now to the superstructure. Here, look for the flatness of surfaces that should be flat such as side tanks and even curves where these might be expected, such as the cab roof. The boiler cleading should not be dented or kinked and should fit well around the bits that protrude through it, like the dome, safety valves and feed clack-valves. This gets tricky on Belpaire fireboxes so watch it here. Joints and edges should be neat, while flush riveting should be invisible and round-head riveting should be evenly-spaced and well lined up. Watch out for the use of inappropriate screw heads such as vastly out-of-scale hexagon heads, or worse still, cheese-heads.

The use of a variety of different screw sizes and heads in platework is a sign of sloppy workmanship, indeed the choice of screws on a locomotive is an indication of quality. Take the fitting of cylinder covers as an example. Scale practice demands the use of studs and nuts. Small hexagon-head screws are not a bad substitute, but round-heads or cheese-heads are a poor choice. Allen screws incidentally are very much 'non-scale' but are acceptable, if not preferable, in high-stress situations such as the fitting of slide-bars, or where largely hidden, such as on frame stretchers.

Next make an assessment of the detailing work on the engine, what LBSC used to call the 'blobs and gadgets'. Things like buffers, handrails, steps, lamps, number and name plates add tremendously to the character of a locomotive, but don't forget that things like this can always be added at a later date on a basically sound but un-frilled engine.

Lastly, there is the question of finish. Bright steel such as the motion-work and buffers should have a fine grained finish, the result of working with fine emery paper, but edges should not be rounded off, nor should the polish be like chrome-plate which would be 'non-scale'. Brass and copper should be polished well, particularly on funnel caps and domes of narrow-gauge and early prototypes. Paint work is a matter of contention. Many say that a high gloss is ideal but personally I have a strong conviction that a dull finish, or egg-shell finish type paint looks far superior on a model locomotive. In either case it should be properly applied, preferably by spraying, and over a coat of a suitable primer, depending on the base metal. There is an excellent range of paints available to model builders and there is no excuse for not getting just the right colour where the engine should be painted as prototype. Lining-out is a feature of many prototype engines and this can be a nice detail on the model if done well, so look at this also.

Right, so, having examined the engine pretty thoroughly, it is time for the steam test down at the local track. No doubt the vendor will take this opportunity to demonstrate the process of raising steam and perhaps some discussion can be made as to whether the deal will include the steam raising blower, driver's trolley, carrying and storage box or whatever. The locomotive should raise steam to 20 lb or so in 5 or 6 minutes, after which the locomotive's own blower should take it to working pressure in a further 5 minutes or so, perhaps more on a larger engine. One safety valve should blow cleanly as working pressure is reached and if it is of the 'pop' type, it will probably give you quite a surprise!

The first few yards of running may well be rather jerky combined with a shower of hot water from the funnel, particularly if the engine has no drain cocks. Thereafter it should even out with a steady pull and even exhaust beats which will blend into a pleasing purr as the train gathers way. At speed, the pull should remain steady and the beats even as the reverser is brought back almost to mid-gear. The mechanics should be almost silent other than the click of the lubricator drive, so once again, listen for any knocking noises which may spell trouble. During the run, look at the rim of the funnel, it should have a coating of black oil, indicating that the lubricator is doing its work. If it is, your face will probably be fairly speckled with oil drops and smuts from the funnel which is the reason why many model drivers wear the so-called 'grease top' hat of the full size engineman.

Throughout the run the boiler should maintain pressure with no problems whilst the pump or injector holds the water level three quarters of the way up the gauge glass. If you have to resort to the handpump at any stage, then something might be amiss. Look round for leaks and weeps while running. The odd drop here and there does not matter but if it runs wet all over the place then watch it. Also watch how the owner deals with the engine after the run. He should drop the fire and blow the boiler down dry, followed by sweeping the boiler tubes and cleaning out the smokebox. Lastly he should clean the whole engine down before putting it away, preferably after it has cooled off completely.

By now you should have a fair idea of the engine and its present owner and be able to make a decision as to whether to buy at his price, bargain a little, or go and look at some other engines. Ultimately remember the choice is yours so don't be smooth talked into buying the wrong thing. The wrong thing may well be too large a locomotive. Somebody will have to heave it on and off the track and if this takes four large men then will you always have help available?

Here then are some of the considerations for those buying an engine complete and ready to go. What about an engine that is essentially complete but not ready to go, that is, in need of attention or restoration? This can be extremely tempting since you can lay hands on an engine for a very low price. I think however that this course of action is to be avoided by beginners in the model locomotive business, indeed it is full of pitfalls for the experienced. Restoration or repair can involve almost as much work as building from scratch and you may end up with something pretty second rate even then. It is often hard to tell what problems lurk beneath a coat of rust and grime on an old engine and very often the answer will be — a lot! A few years ago some friends and I were asked to assess an engine that the amenities department of a certain local council had already bought. They planned to run it on a small railway in a children's park. The sight that greeted us was depressing in the extreme. The locomotive was a free-lance, non-scale design of remarkably ugly lines. It was only $3\frac{1}{2}$ in gauge, which is really too small for this kind of job and was a 4-6-4 tank engine into the bargain which meant there was little weight on the drivers. Worst of all however, was the fact that it was, quite simply, completely worn out, to the extent that new wheels would have been needed, let alone the various bearings and bushes. As to the boiler, no certificate was furnished and most likely that too would have needed renewing. It was, in short, a total disaster, brought about by sheer ignorance combined with not a little bad judgement. We gave our report and the project disappeared from sight. Do not let the same thing happen to you, yours may not be public money.

Before turning to scratch building, there are two further lines of investigation that the beginner can pursue. The first of these is the purchase of a locomotive in kit form. There are several examples of this option now available as advertised in

Don Young, well-known designer of model steam locomotives tends a 5 in-gauge 4-4-0 loco on a ground level track. (Steve Stephens.)

the model press. Kits can vary from a very complete set of parts, right down to the painting, which only need hand tools to assemble, to a collection of machined parts and a boiler. Clearly the latter case will involve considerable work to finish, including all the platework, plumbing and fittings but it will at least halve the time of construction, at a price. This is no bad thing and certainly gets the beginner over all those nasty bits like cylinders and wheels which can cause sleepless nights. One can also purchase many other items complete and ready to fit which is going to be quite a help. These include the pumps, both axle-driven and emergency hand-pump, the lubricator, water-gauge, injectors, check-valves, the various wheel-valves, steam turret, safety valves and even the whistle. These items are described in the catalogues of several suppliers and it is probably a good plan for the newcomer to send off for some of these to browse through in odd moments.

In this section on kits and bits, I might take the opportunity to mention ready-made boilers. Again there are a number of professionals in this field supplying nicely made 'kettles' complete with test certificate at prices varying from £120 for a 'Rob-Roy' boiler, to £800 for a 5 in-gauge 'Britannia' boiler. This is fair value since the cost of materials alone will approach this and quite a bit of hard work is involved in the average boiler. Again this will put the novice well ahead in the completion of an engine.

The more complete kits are a rather different proposition, notable among these are the Japanese OS products. These seem to be well produced, but take quite a time to put together none the less, and a figure of at least 200 hours should be allowed for completion. They are also quite expensive, and the money might be better invested in a complete locomotive. This is very much a matter for the individual to decide.

The final line of investigation before scratch-building is one full of possibilities for the beginner to acquire a good engine at minimum cost in terms of money and effort, but is fraught with the most terrible pitfalls. This is the purchase of a part-finished locomotive. There are frequent adverts for these in the model press and the locomotive dealer may also offer part built engines for sale. These often appear to be bargains, and I believe in many a case they are, but at the same time many will be traps for the unwary. The question one has to ask is: 'Why has the builder given up?'. There will be genuine cases of course, injury, death or whatever, but very often the answer will be that things have gone wrong and the builder has lost heart. You would then be saddled with the tricky work of sorting out his problems. For the beginner, this is rather like starting out with one hand tied behind your back. If the chassis is essentially complete it will probably be advertised as 'air-tested' that is, it has been run on compressed air. You should ask for a demonstration of this, and take the opportunity to listen carefully to the resulting mechanical clatter. If the chassis has not been run on air, it probably won't when you get it home, so investigate further. Be wary about great collections of apparently finished components which are unassembled, because they may not go together properly. Worst of all are part finished boilers. In most cases I suspect a part finished boiler cannot be satisfactorily completed at all, so these are best left out. However, it all comes down to your judgement of the quality of the work you are offered and the reasons put forward for its sale. Most of what has been said about second-hand locomotives can be applied to examining part finished locomotives. Certainly some nice workmanship can be bought for little more than the cost of the castings that have been used.

I have to confess to buying a locomotive in this condition myself not long ago. It was optimistically advertised in the *Model Engineer* as '95 per cent complete', which it certainly was not, but the price was attractive for a spot of speculation. The workmanship was fair and indeed most of the parts were present, but as a partly assembled jumble. I took the gamble anyway. The chassis went together quite nicely and ran reasonably well on compressed air, although I was a little concerned about the amount of play in the axleboxes. These knocked to and fro in their hornblocks as the wheels revolved, but not desperately badly. The boiler also appeared to be moderately well constructed so the various holes were plugged and a water pump fitted. Pressure was brought gently up beyond working pressure of 80 psi and bingo, the firebox crown stays failed and the top of the firebox collapsed. This kind of thing is not really repairable and so the boiler was scrapped. Ah well, you win some, you lose some!

Choice of model for beginners

Having looked at all the options for buying a locomotive ready built, in kit form, or part-built, it seems that the majority of people make the decision to build themselves, not just for reasons of finance, but mainly for the undoubted pleasure of building. So what design should the beginner tackle?

This is a somewhat vexed question, the choice being governed by a number of factors, but I would argue that the main factor should be enthusiasm. Thus, although logic might dictate some particular design, it is perhaps better to tackle the locomotive you really like as you are less likely to fall by the wayside with your favourite engine than with some unglamorous project for beginners. You are going to be working on this project for quite a time and maintaining the initial burst of enthusiasm is sometimes not easy, so better to go for the locomotive you love and you will take delight in seeing it grow under your hands. Otherwise it may become a sort of academic exercise and whilst experience is always useful, most people lack the time to slog through a so-called 'beginner's design' in order to go straight on to the design that they really fancy.

This is not to say that you should start straight on a 5 in-gauge 'King' with its four cylinders and Belpaire tapered boiler, but that you need not 'de rigueur' start with the ubiquitous 'Tich' or 'Rob-Roy'. In fact, although photographs of these locomotives appear in this book, I would counsel against these particular engines for beginners, particularly 'Tich', since I have met a number of people who were disappointed by this outwardly attractive little engine. Two features lure the unwary into the trap, the first being the excellent construction manual *Simple Locomotive Construction — Introducing LBSC's 'Tich'*. This book is a compilation of the very comprehensive set of articles which appeared in the *Model Engineer* in the early fifties, and tackles the subject very thoroughly indeed, for the benefit of the complete beginner. Most of the machining operations are fully described in a light and easy to follow style, sprinkled with anecdotes and advice to leaven the mixture. The book is a must for newcomers and to be thoroughly recommended, even if the locomotive is not. The second lure is the size and apparent simplicity of 'Tich', which seem like good sense but in reality are drawbacks. True, the cost of the castings and materials is minimal but so is the end result. There is really just as much work involved in this engine as something a little larger, and at the end of it all, 'Tich' is just too small. True, it will haul its driver, in fact it has the cylinder power, despite a bore of only $\frac{11}{16}$ in, to pull two

Top right *You can just see the marine-type firebox fitted to 'Sweet Pea' in this picture.*
Centre right *This is the same 'Tich' as featured in Chapter 1, now painted and titled!*
Right *'Sweet Pea' fitted with a tender. A beautiful example of a good beginner's loco.*

people, but the boiler, even the larger revision, is marginal. Whilst a well-built 'Tich' in skilled hands may be capable of non-stop running, the average 'Tich' built by the average beginner tends to make 50 yd hops with long pauses to 'blow up' in between. This is because the little engine has no reserve to allow for not-quite-perfect-construction and heavy-handed driving. The result is a disappointed and frustrated driver. Another point is that although she can show a startling turn of speed, 'Tich' is still slower than most models and tends to hold up the proceedings on club tracks. In the end, the builder usually retires the engine to the mantlepiece, where it looks very pretty, and goes on, somewhat chastened, to something bigger.

Much the same is true of 'Rob-Roy', again offered as a beginner's engine, though with a much bigger boiler than 'Tich' and cylinders of $1\frac{5}{16}$ in bore, this is a more practical proposition and can be seen happily circulating at track events in charge of a good 12 stone. Again there is a construction manual, though nowhere near as comprehensive as the 'Tich' book since it assumes some knowledge of both locomotives and machine work. It is also rather vague in some places and leaves the poor beginner entirely on his own in others, but is more-or-less useful and well worth reading through.

Something similar in shape, but much bigger, is Martin Evans' 'Simplex' design for newcomers. This is a popular free-lance 0-6-0 tank engine and makes a lot of sense. It has easy-to-get-at outside Walschaert's valve-gear with slide valve cylinders of $1\frac{3}{8}$ in bore and general all round simplified design but is still a good solid puller on the track, capable of handling a couple of loaded trolleys. Again there is a construction manual available to aid the interpretation of the drawings.

Another more recent design for beginners by Martin Evans is 'William'. This $3\frac{1}{2}$ in-gauge 2-6-2 tank design makes a great deal of sense for the beginner, being small and simple but none the less both powerful and handsome, it has $1\frac{1}{8}$ in bore cylinders. Do not be put off by the pony trucks, these are simple devices and are identical. They allow a nice long engine accommodating a good sized boiler and reasonably large water tanks. The construction of 'William' was very fully described, especially for the beginner, in a series of articles in *Model Engineer* starting in December 1982.

I personally prefer standard gauge British outline locomotives, but if you like narrow gauge engines such as those which run on the Tal-y-Llyn or Ffestiniog railways of Wales, then there are model versions of these. The advantage here is that for a given gauge, they give a large model of a very simple design and good accessibility. Thus Martin Evans' locomotive 'Conway', though a $3\frac{1}{2}$ in-gauge 0-4-0 tank, has cylinders of $1\frac{3}{16}$ in bore and is large enough to haul a couple of

Top right *Cab view of* Lady Caroline. *The cutaway in the bunker allows coaling of the fire.*
Centre right *A very nice example of 'Rob-Roy' in steam and ready to go. The rod protruding through the left-hand water tank is the hand pump handle. Note the lubricator mounted on the running board and driven by the crosshead.*
Right *This 'Simplex' is also waiting for the signals to change. It would be quite happy to set off with half a ton in tow, not just its driver.*

Above left *Waiting for the signals. 'Rob-Roy' can certainly pull a live passenger. Note the driver's eye protection; this is a wise precaution because a hard working engine throws plenty of gritty ash up its funnel.*
Above right *The one big advantage of 'Tich', portability!*

trolleys. It was described in the *Model Engineer* starting in April 1980. 'Sweet Pea' by Jack Buckley is a similar sort of machine for 5 in gauge and features simple Hackworth valve gear and a marine-type boiler with a circular firebox. This attractive engine is proving very popular and is a good choice for a first go, having been described very comprehensively in a series of articles in the magazine *Engineering in Miniature*.

Another good choice could be the locomotive called 'Marie E', a slightly quaint early American 0–4–0 with a four-wheel tender, by Don Young. This 5 in-gauge engine is again very simple, with slide-valve cylinders, with the port face on top like 'Tich' and 'Simplex'. This arrangement makes for the easiest possible assembly, particularly in regard to valve-timings, which is a little tricky on 'Rob-Roy' with its inside steam-chests and valve-gear. By way of a comparison, 'Marie E' has a cylinder bore of $1\frac{1}{8}$ in.

Similar to 'Marie E' but of British outline is another Don Young design, 'Rail Motor', based on the engines of the Rhymney Railway of South Wales. Though perhaps a slightly odd design to some eyes, the so-called 'Number 2' version of this 0-4-0 locomotive is designed especially for the beginner. It has a modest but adequate cylinder bore of 1 in and features outside Walschaert's valve-gear.

Of course, if you are not satisfied with any of the designs available you can always set up a drawing board and produce your own. There are quite a number of books to help you, starting with Henry Greenly's early book *Model Steam Locomotives* and going on to more modern works such as Martin Evans' *The Model Steam Locomotive — A Complete Treatise on Design and Construction*. Read these carefully and look through some books on full size locomotives to choose a prototype, or draw up a free-lance design to suit your tastes, but do not forget that suitable castings will probably have to be provided for the model, at least for wheels and cylinders. If existing castings cannot be pressed into service then you will have to make your own. There are various ways of going about this. First and most expensive, simply give the drawings to a foundry who will make up patterns and cast the items in whatever metal you choose. Second choice is to make up patterns yourself and have them cast by a foundry. Pattern making is an art in its own right but armed with the information from a book such as *Foundry Work for the Amateur* by B. Terry Aspin, the enterprising model maker can achieve surprisingly good results. This leads to the third option which is to make patterns and then to do the casting also yourself. Large items such as wheels and cylinders are generally cast in classic split sand moulds, but for small and detailed work investment, or lost wax, casting may be worth a try.

There is no doubt that although very time-consuming, to do your own design work, casting and building is ultimately rewarding for the model locomotive enthusiast. This was a course chosen by two complete beginner friends of mine some years ago. The resulting $7\frac{1}{4}$ in gauge 2-4-0 locomotive *Edwin* runs beautifully around their large garden railway and has the finest chime whistle I have heard!

A word of warning here for beginners about the general quality of drawings and castings provided for model steam locomotives. If you come from an engineering background and are used to correctly drawn and toleranced plans to work from, then you will find things rather different here. Few tolerances are given, instead the terms 'full' and 'bare' are applied to dimensions to indicate the sort of fit required. As an example, take the knuckle joint on a coupling rod. (See Chapter 5.) The slot in the fork might be dimensioned 'full $\frac{1}{4}$ in' whilst the tongue designed to fit it would be dimensioned 'bare $\frac{1}{4}$ in'. This is a way of indicating that the parts should fit closely, but with enough clearance to permit free working.

By and large you are expected to use common sense to make one piece fit another, though legends such as 'ream to fit axle' do appear. Indeed quite a lot is left to the interpretation of the builder, particularly on earlier designs. This is no

Above *'Marie E', quaint but quite a good proposition for the newcomer, capable of hauling a fair load.*

Below *Here is another example of 'Marie E' pictured from an unusual angle.* (Patrick Advertising Associates.)

Above *Rather a jolly little engine, here is a good example of Don Young's design 'Rail Motor' on a ground level track.* (Patrick Advertising Associates.)

Below *'Rail Motor' and scale rolling stock pictured from a realistic ground level angle.* (Patrick Advertising Associates.)

bad thing really, but something you should be aware of. Rather more serious are some of the errors and oversights which inevitably creep into design work. There does not seem to be a system for correcting these so it is a good plan to seek out someone who has built your intended design, buy them a drink or two, and quiz them carefully about the construction of the engine. Mind you, model drawings are, one has to admit, very cheap considering the work involved.

Castings too can sometimes leave a lot to be desired. Clearly from time to time, the odd blowhole will appear in castings from the finest supplier and he will generally replace the faulty item without quibble. However it pays to examine castings carefully before starting and reject any that are obviously faulty at once. Here again, seek out the advice of someone who has used your intended casting supplier and see what he says.

The majority of castings offered are still made by sand moulding and I personally feel that changes here are long overdue. Processes such as investment casting are well known in industry and though relatively expensive to tool-up, produce excellent results. Indeed fine castings might eliminate much lengthy and arduous matching; imagine cylinder castings for which the total work was to run a reamer through bore, rub the working faces smooth on fine emery cloth on a surface plate, and tap the stud holes! Better materials and drawings are creeping into the hobby, but the process is a slow one. So, here are some ideas to think over before deciding what you are going to tackle. What design do you like? What will your workshop handle? What can you afford? How skilled are you at working metal in general? How much time can you devote to the project? What help is available to you in the way of advice or machining facilities for large components? What can you lift around the workshop without becoming a medical statistic?

Not a bad choice for the beginner, this early single-wheeler has the absolute minimum of complication. In 5 in-gauge, it puts up a very creditable performance and gives a scope for plenty of Victorian ornamentation.

Above *A selection of wooden casting patterns made by the Holt brothers for their 7¼ in-gauge tank engine project.*

Below *An iron casting and a completed cylinder block for* Edwin.

Above Edwin *begins to take shape on the workshop bench. All the machining work for this engine was done on a Myford Super 7 lathe.*

Below *Close-up showing the slide valves fitted to* Edwin.

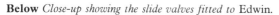

Above Edwin *complete and ready for the road. Fine work by two complete novices with no engineering training or experience.*

Below *Dunstan Holt designed these pistons so that he could fit twin piston-rings without the risk of breaking them during fitting. The cylinder cover has a milled depression to aid steam transfer.*

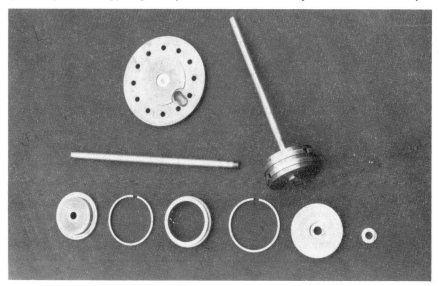

A little unconventional but utterly practical, this loco is ideal for the garden railway. The side tanks actually carry coal whilst water is carried in the driver's truck.

What do the rest of the family reckon? These are the considerations, think about them and then take a trip to one of the model engineer's suppliers, buy the drawings, the construction manual, if there is one, and make a start.

Summary of designs for beginners

Name	Wheel arrangement	Type	Gauge	Cylinder bore	Valve gear	Remarks
Tich	0–4–0	Side tank	3½ in (5 in and 7¼ in available)	$^{11}/_{16}$ in	Outside Walschaert's or slip-eccentric	Very small, free-lance design
Rob-Roy	0–6–0	Side tank	3½ in	$^{15}/_{16}$ in	Stephenson's inside	Small
Simplex	0–6–0	Side tank	5 in	1½ in	Outside Walschaert's	Good sized popular loco, free-lance design

Above *Three different versions of Tich in a row. The first is a large-boilered version with slip eccentric valve-gear, the second a small-boilered version with slip eccentrics and the third is the classic small-boilered version with Walschaert's valve gear.* (Bill Mawson)

Below *A beautiful 5 in-gauge Adams 4-4-2 tank basking in bright August sun. How would you like this on your mantlepiece?*

Above *A very nice 'Princess Marina' designed by LBSC for 3½ in gauge. Not a bad choice for the beginner but note the Belpaire firebox.*

Below *This is a summer scene on Dunstan Holt's garden railway with* Edwin *in charge of a two-coach train.*

Above Edwin *in detail, not a scale model and some would say not pretty, but a thoroughly practical design by two complete beginners.*

Below *This however, is a scale model and a very fine one too. Look at the detail on this 5 in-gauge nineteenth century 2-4-0 engine. I have driven her a couple of times and she runs superbly.*

Left *Big stuff now at the Dobwalls Miniature Railway in Cornwall. This 7¼ in-gauge American loco hauls heavy passenger trains for a living. The railway is well worth a visit if you are in the area, but the loco is a bit heavy for beginners.*

Below *Small is beautiful. This is 'Fayette' a 2½ in-gauge 'Pacific' designed by LBSC. The polished metal looks pretty but I prefer to see a proper paint job myself.*

Name	Wheel arrangement	Type	Gauge	Cylinder bore	Valve gear	Remarks
William	2-6-2	Side tank	$3\frac{1}{2}$ in	$1\frac{1}{8}$ in	Outside Walschaert's	Excellent design for beginners, free-lance but scalish
Conway	0-4-0	Saddle tank, narrow gauge	$3\frac{1}{2}$ in	$\frac{13}{16}$ in	Outside Hackworth	Simple but large
Sweet Pea	0-4-0	Saddle tank, narrow gauge	5 in	$1\frac{1}{2}$ in	Outside Hackworth	Simple design and good puller — good beginner's design
Marie E	0-4-0	Four-wheel tender	5 in	$1\frac{1}{8}$ in	Inside Stephenson's	Quaint American design, but good for beginners
Rail Motor No. 2	0-4-0	Four-wheel tender	5 in	$1\frac{1}{4}$ in	Outside Walschaert's	Again quaint but British design
Juliet	0-4-0	Side tank	$3\frac{1}{2}$ in	1 in	Slip-eccentric, inside Stephenson's or outside Baker	Nearly as big as Rob-Roy and simpler

Chapter 4

What you need

Given the necessary means, one could spend a huge sum setting up a workshop for the construction of live steam model locomotives, indeed this is a rather jolly piece of conjecture for idle moments. A three-roomed brick structure with central heating and air conditioning to house drawing office, boiler shop and machine shop would fit the bill. Machine tools would include lathes of 5 in and $3\frac{1}{2}$ in centre height and vertical and horizontal milling machines, all equipped with modern digital scale readout. Then there would be large and small bench drills, power hack-saw, band saw, guillotine, bending-rolls, fly-press and so forth, not to mention a wide selection of hand tools. Total cost? Not a lot of change from £100,000 and probably not a lot of satisfaction either.

Much better to go the way of most people and produce a good model with the minimum of tooling. You can actually make a start and progress a fair way with only a few hand tools, then gradually build up your equipment, begging or buying items as you need them. Once again contact with the local model club is invaluable here since members will often pass on tools and machines for one reason or another, and most clubs hold annual bring and buy sales which are a great source of bargains. Investigate the machinery column of the local newspaper and keep an eye on the classified advertisements in the model press to gain an idea of what prices are like and what is available in the way of workshop equipment.

The workshop
One obvious must is a place to work in peace and quiet and safety. Models can be built on the kitchen table but having to clear away each evening and keeping the children off those nice bright, shiny pieces of metal makes for rather slow progress. It is really vital to find a suitable place to closet yourself in comfort. A room in the house is the best, if your abode is large enough to accommodate it, failing this a brick-built extension to the house is a worthwhile proposition, since it will undoubtedly add to the value of the property. Next plan is a shed in the garden, but beware here of two problems. One is heating since the workshop must be dry and at least reasonably warm to prevent the machines from rusting. It should also be warm for the comfort of the inhabitant, otherwise, come the cold winter nights, enthusiasm will wane, the short-term dubious pleasures of the television will hold sway and progress on the locomotive will cease; maybe never

This is a view of part of my workshop. It is pretty small but at least everything is to hand.

to start again. So, a garden shed is fine but it will need careful attention to weather-proofing, excellent insulation added and some form of heating installed. It will, of course, need a mains power supply to run lights and tools and this is something to consider when siting the workshop. Power is not generally a problem if you decide to use the garage, but again, insulation and heating will be needed.

The second problem is that of security in these days of burglary and vandalism. It depends where you live, the inner cities being the real problem, but bear in mind that a fully equipped workshop is valuable and prone to mindless damage which can be quite heart-breaking, hence the preference for a room in the house. This leads on to a further consideration, that of noise. Many operations, such as hack-sawing and riveting are pretty noisy and the running of a lathe can produce a penetrating rumble; so think of the neighbours before pushing the lathe up against a party wall. Weight can also be a problem, because machine tools are pretty hefty and need a firm even base to run accurately. This tends to rule out upstairs rooms, particularly if the lathe is larger than 3 in or $3\frac{1}{2}$ in centre height.

Basic equipment

So, what do you need first of all to make a start? A good solid workbench is the primary requirement, heavy enough to take a fair amount of punishment from energetic sawing and filing and fitted at the right height for you to work at. It really must be fixed either to the floor or walls, because nothing is more aggravating than a bench which wanders across the floor whilst you are operating on some piece of metal. Fixing also helps to prevent annoying 'chatter' when filing and sawing. As to the actual worksurface, many people use plain wood, which does absorb shocks and lessen damage to tools and metal components as they get 'shunted' around the bench, but it also absorbs oil. In time it becomes soft, black and scarred which is rather unappealing, so I prefer a formica type finish which is hard and easy to clean down. Things do tend to slide across laminate tops but this can be an advantage when moving a hefty locomotive around during construction. A second bench is always useful, indeed horizontal space of any kind will get well used. It is suggested that the second bench be designated 'clean' for drawings, marking out and assembly work in which case it might well be set lower to allow for seated operation. This is one of the points of comfort which in the long term will ease the project through to completion.

Next you will need a bench vice, or again, preferably two. One should be fairly small, say about 3 in jaw size, the other a bit larger for the heavier items, say about 5 in or 6 in jaw. In both cases the serrated jaw-faces supplied should be removed immediately and replaced with either brass or alloy ones to prevent damage to the faces of work pieces. The fibre jaw fittings that the vice people supply are okay, but tend to be too soft and inclined to fall off the vice in time. Like many engineering items, you will find that there are vices available from iron curtain and far eastern countries costing a fraction of the price of their British counterparts. You will have to decide if the quality is adequate for the purpose, but there is one point to watch here. Make sure that on the vices you buy, the back jaw will be positioned far enough forward to overhang the bench. If not, then it can only hold items above bench height which means you cannot work on the ends of long components like main frames. This is very limiting and annoying so do be careful.

Next items required — basic marking-out tools. A good quality steel rule is priority here, with five clear divisions and a satin finish. A 6 in one will cover most jobs and fit nicely in the top pocket. Then you need a scriber, here a tungsten-carbide tip will save time, keeping the tip sharp, but a conventional steel one is fine. For centre punching or popping I like the click-action auto-punch but a piece of silver steel (see last part of this chapter) suitably sharpened with a 90° angle and hardened, will do the same job. A good set-square is essential and again, one of about 6 in size will cover most eventualities. The properly equipped engineer's shop will no doubt have a range of sizes of most hand-tools, such as set-squares and rulers but the beginner can get by quite happily with just one item most of the time. A pair of dividers for laying out circles completes the beginner's marking-out kit for a cost of a few pounds. Once again, the proper engineering works will have a selection of surface tables, height gauges, great collections of odd-shaped callipers and so forth, but these are not necessarily for us. The only other item for the kit would be a bottle of marking blue, which is a dye that is painted on to metal surfaces so that the bright scribed lines show clearly through. Personally I do not find much use for the stuff, but this is just a question of preference.

We can now mark out our work so we need some tools to cut it with. The prime tool here is the hack-saw which is likely to do quite a lot of basic work so it pays to get a good comfortable one with a pistol type grip, to take 12 in blades. Do not bother with the old black carbon steel blades but go for the blue high-speed steel ones, about 32 teeth per inch, for most jobs. These are more expensive but last far longer, especially if used with some lubricant when working on steel. For smaller work, the fine-toothed 6 in 'junior' hack-saw is very useful and a cheap item to acquire. Equally hardy is another hack-saw frame fitted with an 'abrafile' blade, though it is not essential. Files are next down the line and again, these will do a lot of work, believe me, so it pays to develop a good file policy. Rule number one of this policy is very simple, it is to throw away worn files. Let me repeat the instruction — files which are past their cutting prime should be placed, forthwith, in the dustbin. Do not pretend that you are going to get them reconditioned by acid dip, or that they can be turned into useful scrapers, just ditch them. Old files are a menace, so, if you inherit a drawer full of black, dirty files from some ancestor, remove the handles and liberate the drawer space for something useful. This brings in point number two, always fit handles to files, even the small ones. Buy a selection of handles at the start, they are very cheap and make the file safer and easier to use. Next point, store them carefully. If they are allowed to jumble together in a heap they will quickly damage one another and become useless. I keep mine resting on the top of a cupboard at hip height which means that they can be drawn out smartly for use and returned afterwards. You will need a selection of files but to begin with perhaps half a dozen different ones will suffice. These items are rather charmingly categorized by coarseness of tooth; 'bastard', 'second cut' and 'smooth'. A suggested selection might be made up of:-

12 in bastard flat file
8 in second cut flat file
8 in second cut half-round file
6 in second cut square file
8 in second cut round file
4 in second cut warding file.

Added to this a set of little needle files is very useful, but here one can buy a file to suit the job in hand rather than lashing out on a complete set at the start.

The same philosophy is true of taps and dies, which can be an enormous expense to buy as complete sets, particularly in high-speed steel. The answer here is to pick up the taps and dies you need as the various jobs come up, thus spreading the load. Eventually a complete set will build up but you will find that a few sizes will cover much of the work on an engine. The two types generally specified are BA (British Association) for the smaller threads and ME (Model Engineer) for the larger. At one time we were threatened with metrication but this plan, having been partly enacted in the engineering world, has now receded again. Thus, I think that the beginner can quite safely commit himself to imperial standard equipment, though I personally would welcome metrication and would happily junk my imperial gear to change to the vastly superior metric system.

BA sizes run from 0BA down to the incredibly small 16BA, the use of which entails extreme skill or a pact with the devil. Invest in sizes 2, 4, 6 and 8 and you should cover most eventualities, though the odd sizes 5 and 7 do come up quite often, especially in old LBSC's designs. You can, of course, change sizes and say use 6 instead of 7 but take care with your choice in this case. ME threads are more varied, having a choice of pitches normally 32 or forty threads per inch, with a range from $\frac{1}{2}$ in × 32 threads per inch to $\frac{1}{8}$ in × forty threads per inch.

Look over the drawings of your chosen locomotive and see which threads are specified and purchase these as you need them. Do not go in for second-cut taps, the taper will do most jobs and you can get the plug tap if you have to get to the bottom of a blind or stepped hole, as with, say, piston rod glands. Unless you are a bloated capitalist or have won the pools, don't get the expensive high-speed steel taps and dies but settle for carbon steel. By and large these tools will get relatively little use and most of that will be on easy going brass, so use carbon steel, use it carefully and you will have no problems.

Incidentally, a little tip here; should you have the misfortune to break a tap deep in some complex workpiece such as a cylinder block, don't throw yourself off Beachy Head, but throw yourself into the *Yellow Pages* and find a local engineering company with a small spark-eroder. This cunning machine can erode the hard metal of a tap whilst leaving the softer metal of the workpiece unscathed, not even damaging the threads already cut. For the price of a pint or twenty cigarettes one can retrieve a potential disaster, so bear this in mind.

You will, of course, require tap wrenches and die stocks in order to use the taps and dies which you purchase. In this case the cheap far eastern varieties are probably adequate, or else you could scavenge through those second-hand tool stalls to be found in many a market place. These can be the source of many treasures and bargains and, for example, a second-hand hacksaw fitted with a new blade is just as good as a brand new one.

Now, what about some absolutely vital equipment — measuring tools. We have already talked about rulers, but what about the high accuracy stuff? The classic solution is the micrometer for dimensions up to 1 in and a vernier calliper gauge thereafter, indeed, this combination works well and is to be found in many a workshop throughout the land. However, times have changed, technology has rolled on and much better things are available to us now. The problem with the micrometer, and even more so, the vernier, is that they are difficult to read and demand all sorts of mental gymnastics to make sense of the results, leading to frustration and error. Time served traditionalists may scoff here but my statement is true. The tool to go for now is the dial-reading calliper gauge. This will give inside and outside diameters and depth, between zero and 6 in, precise to a thou (one thousandth of an inch) and very easily read. These things cost around £40 but this is well spent since a micrometer and vernier are not a lot less, and you will save a vast amount of time and effort on the large number of measurements that need to be performed whilst building a locomotive. Better still, if you can run to it, is a calliper gauge with a digital readout. These cost around £80 but give direct readouts to four decimal places, switchable between imperial and metric, and can

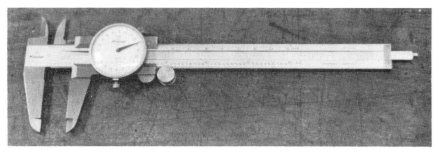

Above *Calliper gauge fitted with a clock dial. An extremely useful measuring instrument.*
Below *Probably the ultimate measuring tool for the model engineer, a calliper gauge with digital readout.*

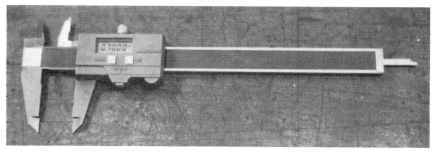

be zeroed at any position to give immediate offsets without arithmetic. These are absolutely wonderful, but a little beyond the means of the average man. Whichever you choose, look after the instrument very, very carefully, particularly the sharp points of the inside diameter side. Keep it clean and put it away after use, and you will have a faithful friend for life.

What other hand tools are needed? A range of odds and ends, screwdrivers, small spanners, pliers of various sorts, allen keys, hammers large, small and soft-faced, hand-drill and so forth, all of which can be acquired as you go along, whilst things like toolmaker's clamps can also be made up in the workshop.

The lathe

You can certainly go a fair way in the construction of an engine with just hand tools, as far as erecting frames, stretchers and buffer-beams, but pretty soon you are going to need that most ubiquitous of machine tools, the lathe. Equipped with a modest range of accessories, this beasty can tackle a huge variety of machining tasks, from simple turning to facing, milling, gear cutting and so on. The lathe will probably be the largest single investment for the newcomer, so it pays to take care in the choice of this machine. Indeed it would be best here to refer the reader to some of the more specific books on the subject (see chapter 11) but let me say just a few things on the subject.

First of all the matter of size of lathe, this being gauged by the centre height, that is, the distance from the turning axis of the machine to the bed on which the saddle slides. Thus twice centre height is the largest diameter object which the lathe can handle, except in the case of gap-bed lathes, where short workpieces of a larger diameter can be accommodated. Twice the centre height is called the swing of the lathe, a term used by our American colleagues to gauge lathe size.

There are several miniature lathes on the market, some quite sophisticated, with centre heights of around 2 in to tempt the novice. These are cheap and compact, and would appear to be able to handle the components of a small engine like 'Tich', but beware, they are just not man-enough for the task. For our kind of work we really need a machine of more than 3 in centre height with a good size mandrel bore (the hole down the main shaft), power feed to the saddle, micrometer dials and a wide range of accessories for maximum versatility. One make in particular is outstanding in the kind of features required by the model

Above *One of the most popular lathes among model engineers, the Myford ML7. This is my own venerable example.*

Below *The EMCO compact 8 lathe can provide good turning capacity in a small workshop space. This machine will cost you something around £1,000 new.* (Elliot Machine Tool Co.)

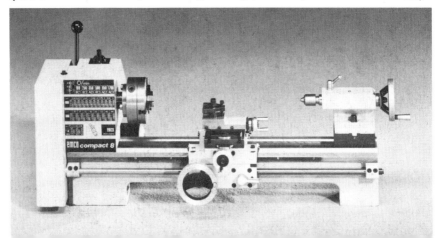

engineer and hence has become almost universal in home workshops. This is the Myford company of Nottingham, whose ML7 lathe of $3\frac{1}{2}$ in centre height is the nearest thing to standard equipment in our hobby. Not only is there a large selection of tools for the ML7, but many articles have appeared in the model press with designs for further clever devices to extend the use of this already versatile machine. Slightly more sophisticated is the Super 7 with indexable micrometer dials, power cross-feed and better mandrel bearings allowing a wider range of speeds. A useful feature of these lathes is the gap-bed which will accommodate some pretty large workpieces. A similar range of machines is offered by Boxford of Halifax. These are a little larger, at 4 in centre height, and although they do not

A very nice machine for the model loco man; the Boxford AUD lathe, complete with tray, stand, light and chuck guard. (Boxford Lathes.)

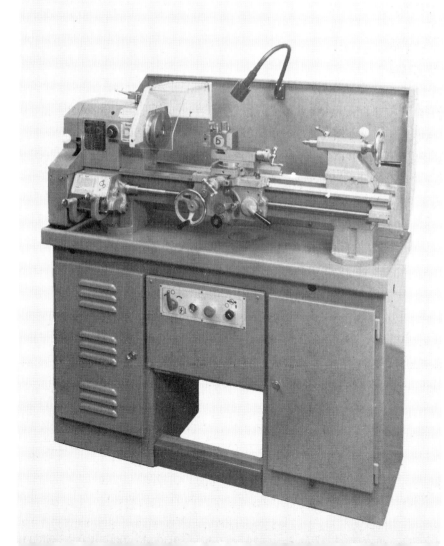

have a gap-bed they are favoured by some engineers as being more rigid. As with many things the ideal course of action is to buy brand new, but the cost of a new Myford or Boxford may well cause the beginner some headaches, even more so when the cost of a few accessories is added to the price of the basic machine. At the time of writing, the basic Myford ML7R (the latest model in the ML7 range) without motor, chucks or stand, costs about £1,000 and the Super 7, similarly equipped, £1,200. The recent addition to Myford's range, the 254S comes with a ¾ hp motor and a three-jaw chuck, but the price goes up to around £2,000. The Boxford AUD comes complete with motor, switchgear, stand and chuck at about £2,500, however, this model is being phased out in favour of Boxford's new X10 range. These have even higher specification but naturally cost more, starting at around £2,750.

What can be done to avoid this sort of hefty investment? There are two obvious answers, buy a second-hand machine or buy one of the much cheaper far eastern products that are much advertised in the model press. These oriental manufacturers produce a whole range of tools from lathes, milling machines and bench drills, down to machine vices and clamping kits, all at much lower prices than one would expect to pay for the equivalent British equipment. The reason for this can only be that the all-round quality is lower, the problem is to decide if the quality of the main functional components is good enough to do the job. Certainly there has been a catalogue of horror stories in the letters' columns of the model magazines but I have seen some of these machines perform perfectly well. The British importers reckon to carry out assembly and quality checks in order to maintain standards and the market for foreign machines continues to grow, no doubt at the expense of the British machine tool industry. It is not really possible to generalize, only to suggest that you examine a Taiwanese lathe very carefully and ask the salesman what guarantees he offers.

One advantage of these machines is that they very often come fully equipped with chucks, motor, stand, gearbox and even coolant pump. By contrast, the traditional British machine is supplied in very basic form and there is considerable further expense involved to equip it sensibly. Quite a few lathe accessories will be unnecessary for the beginner, things like coolant pumps or taper turning gear are just frills for us, but one idea that I find very useful is the quick-change toolpost. Here the lathe tool is held in a special block, which itself has a quick-clamp device on to the toolpost. This means you can have a range of tools, each ready in one of the special blocks and thus change tools very quickly without the encumbrance and danger of the four-tool turret or back toolpost. This system, combined with the use of a keyless tailstock chuck and a good dial-reading calliper gauge, will make you into a demon turner, producing accurate components, easily and quickly, by the boxful!

For myself, I believe that going for a second-hand lathe is a very good course of action for several reasons, not least of which is that it will probably come complete with motor, chucks and probably several boxes of odds and ends accumulated over the years. The classified advertisements in the model magazines will give you a good idea of what is available and the sort of price, *Exchange & Mart* is also worth a look, but broadly speaking a beginner could buy a perfectly adequate lathe for between £250 and £500. Below this figure, you might expect to do some renovation, which is perhaps a good thing in that you would become more familiar with the workings of the machine. Several advertisers in the *Model Engineer* offer a reconditioning service for lathes, including regrinding of the main slideways, for something around £150 or £200. Thus you might buy an old or much-used machine and have the main working parts attended to by a professional, clean and paint the minor parts yourself, and end up with a lathe in practically new condition. Take care with the installation as described in the lathe books and you will have another friend for life.

Instructions for the purchase of a second-hand lathe are really beyond the scope of this book and the reader should refer to some of the more specialized publications (see Chapter 11). Once again the local model engineering club can probably give plenty of helpful advice and if you can persuade someone with experience of lathes to go with you to look at a second-hand lathe, so much the better. Scan the machinery column of the classified ads in your local paper and try phoning local machinery dealers to see what is on offer.

By way of a summary, here is a table of some of the various accessories that can be fitted to or used on most lathes, together with some indication of their usefulness in model locomotive building.

Accessory	Remarks (In relation to beginners)
Face-plate	Vital
Three-jaw chuck	Vital
Four-jaw chuck	Vital
Tailstock or drill chuck	Vital
Backgear	Useful
Self–act or auto feed to saddle	Very useful
Self–act or auto feed to cross-slide	Handy
Indexable micrometer collars	Very useful
Vertical slide	Vital if you have no milling machine. Needs a machine vice
Four-tool turret	Better than nothing but prefer quick-change tools
Back toolpost	Useful for parting-off only
Tailstock die holder	Very useful
Taper turning gear	Rarely used
Screwcutting gear	Rarely needed, mostly used as a self-act for which screwcutting feedrates are too high. Fit fine feed gears if available.
Collets	Handy, buy only the sizes you need, not an expensive set.

Angle plate	Very handy and vital if you have no milling machine. The Keats variety is probably the best
Fixed steady	Handy
Travelling steady	Not often useful
Centre drills	Vital
Parting tool holder	Vital
Boring bars and fly cutters	Vital, but you can often make your own
Knurling tool	Something to borrow from a friend
Dividing head	Nice, but expensive, the beginner can get round this one
Rotary table	Same thing
Spherical turning gear	Don't bother
Drilling and milling spindle	Handy but not essential
Filing rest	Very useful, make one
Special milling attachments	Okay but better to go for a real vertical milling machine
Slitting saw mandrel	Handy, but better to have a miller
Quick-change tool-holders	A wonderful boon, very useful
Carbide-tipped lathe tools	Extremely useful, last ages, vital for cast iron. Throw away when blunt
Dial test indicator on magnetic base	Expensive but pretty useful
Clamping gear, dogs, tee-nuts, etc	Vital, collect these as you go along
Rack feed tailstock	Very nice but not essential
Machine vice for vertical slide	Vital for small milling operations
Digital readout	Suitable for daydreamers, this costs around £1,000 per axis!

Other machine tools

Having acquired a suitable lathe, what other machines will be most useful in getting your pride and joy into steam? The first choice here must be a bench drill, because, although one can use the lathe to drill accurately, it is both difficult and rather dangerous, and there is a great deal of drilling to be done on a model steam locomotive. The drill should preferably be of $\frac{1}{2}$ in capacity with a good chuck to accept very small drills, and a range of several speeds. As with lathes, a British machine, such as Fobco, or Startrite (priced about £300) is desirable, but an oriental machine is probably adequate. Something that is not adequate however, is an electric hand-drill mounted in a flimsy die-cast zinc drill-stand. These tend to be inaccurate, flexible and very noisy.

Bolt the drill down to the bench firmly, in a position with at least a couple of feet of space either side in order to accommodate long workpieces, such as locomotive main frames, and tie the chuck key to it so that it does not wander away mysteriously.

The next machine you are going to need is a grinding wheel to produce lathe

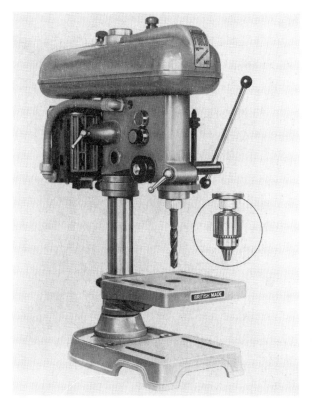

This bench drill by Fobco will make light work of most of the drilling on a model loco. (F. O'Brian and Co Ltd.)

tools and sharpen a variety of other tools such as drills, D-bits and so forth. Nothing particularly fancy is necessary, if you really must sharpen complex milling cutters, then visit the local engineering works. A double-ended spindle with a coarse wheel on one end and a fine one on the other makes sense, and adjustable tables or rests are very desirable. A relatively cheap foreign machine, at about £40, will probably do this job okay, or you can make up a grinder using a commercial spindle and bearing and a suitable electric motor with a V-belt drive. For fine finishing, a good oilstone will also be needed, to complement the grinder.

Having obtained all these bits of kit, your workshop is fairly well set up, but there is one more machine which, though expensive, is a wonderful time and effort saver and well worth some extra work to obtain. This is a milling machine. As with drilling, the lathe can be adapted to perform most milling operations by judicious use of a vertical slide mounted in place of the top slide on the saddle. This means that a workpiece can be moved in all three axes, relative to a suitable cutter held in the three-jaw chuck or preferably a collet. (A collet is a one-size screw-down device for firm and accurate chucking.) The cross-slide will move the pieces from front to back, the vertical slide takes it up and down and moving the saddle takes it to left and right. To move the saddle by small increments usually means engaging the leadscrew and turning this by hand. The leadscrew needs to be fitted with an indexed wheel and the self-cut gear-train needs to be disconnected or you will be continuously back-driving a heavy set of gears. All of this takes a fair time to set up, and, even when this is done, the result is not really rigid enough, leading to slow work rates and possible chatter problems. Also, it is pretty difficult to see what is going on since you have to peer round over the headstock to look at the working face of the job.

Far better then to get a proper milling machine with a horizontal table on which the workpiece can be directly clamped, or held in a machine vice. This table can then be moved left to right, front to back and upward or downward using sensibly placed indexed handwheels. The milling cutter may be mounted on a horizontal shaft running from front to the back of the machine in which case it is, not surprisingly, called a horizontal milling machine. Alternatively, the milling cutter may be mounted on a vertical shaft, hence, vertical miller, and this beast is really rather more useful to us. If the vertical shaft is mounted on an assembly which can be moved around various axes, we have an even more useful machine called a turret mill. This set-up is perhaps the ideal, but tends to be both large and expensive. All we need is a simple, basic vertical mill, although power feed, at least to the left and right motion, is very desirable.

You will find this tool extremely useful for producing recesses and profiles in many locomotive components, such as hornblocks, axleboxes, cylinder blocks, coupling rods, valve-gear and a whole host of other bits and pieces. A lot of these jobs can conceivably be done using the humble file, but a miller is both more accurate and much quicker. It might even be said that a miller is more useful than a lathe, there being more millable operations on a locomotive than turning operations. It would not be unreasonable for a beginner to set up with a good vertical mill and just a small lathe for making items such as boiler fittings. The larger turning operations, of which there are not many, could be farmed out or done on a borrowed machine.

There is quite a variety of milling machines on the market at present and, of course, the inscrutable orientals feature strongly here with some tempting looking bargains, as with lathes and drills.

The little 'Amolco' mill, as advertised in the model press, is very handy, reasonably priced and British. It is, however, somewhat limited in capacity. This machine is a lathe attachment, fitting in place of the tailstock, and using the saddle and cross-slide as a milling table. This is not a bad plan, certainly it is better than using the vertical slide on the lathe, but, once again, the lathe is tied up by the milling arrangements and time will be spent shunting the attachment to and fro. It is, of course, fairly heavy which is an important consideration for the less fit locomotive builder.

Next up the scale comes a miller designed by a model engineer (the late Edgar Westbury) for model engineers, the Dore-Westbury miller. This can be found as a simple machine, or in more sophisticated form with power feed and so forth. Interestingly, for those setting up a workshop on a budget, the Dore-Westbury is

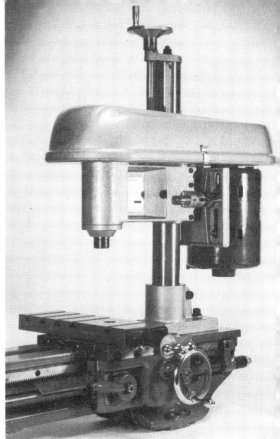

Above left *The EMCO FB2 mill, a versatile and compact miller which would suit the needs of model locomotive building very well.* (Elliot Machine Tool Co.)

Above right *A cheap solution to the milling problem. The Amolco milling attachment, mounted on a Myford lathe. This little machine costs in the region of £250.* (N. Mole and Co Ltd.)

available from Model Engineering Service of Chesterfield in the form of a kit which can be completed on a 3½ in lathe, all the heavy machinery being done already. The cost is around £500 in basic form and £800 in de-luxe form. Although this seems like a good idea, be warned, you may find yourself generally side-tracked into building workshop equipment and tools, at the expense of the locomotive which was your original target.

Similar to the Dore-Westbury, but more in the Rolls-Royce line is the 'Type E' mill from Tom Senior Ltd (now owned by Denford Machine Tools of Brighouse, Yorkshire) costing perhaps £1,800. Beyond this is the 'Junior' model from the same company which can provide for vertical milling, horizontal milling and drilling, but this is perhaps an 'overkill' for the beginner, particularly with a price around £2,400. Another handy looking machine is the FB2 mill from EME Ltd of London, which, at £1,300, bears consideration.

Investigation of second-hand machinery dealers will probably present you with

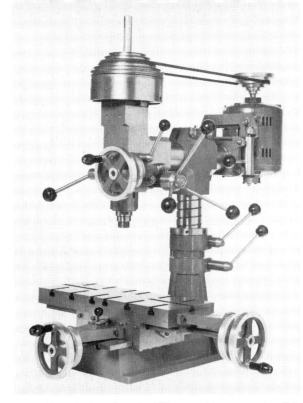

Designed by a model engineer for model engineers, the Dore-Westbury miller is a popular and economical choice. It can be fitted with power feeds and belt guard as required. (Model Engineering Services.)

a startling choice of used milling machines, mostly of the larger variety, with tables around 3 ft or 4 ft in length. This is the result of the recent recession in the engineering industry which has sent many conventionally equipped engineering firms to the wall. Those that have survived the economic blitz have retooled with modern CNC (computer numerically controlled) machines, thus there is a glut of used equipment. The problem here is that the machines are too large for the average home workshop, and many will require a three-phase power supply.

Once again, the classified advertisements in the model magazine are a good source of information on the second-hand market, but at the time of writing I should suggest the beginner pay not more than say £500 for a second-hand small milling machine, and that should include things like a machine vice and auto-lock chuck to take the milling cutters.

Drills, reamers, cutters, etc

Having discussed hand tools and the major machine tools, what else do you need along the way to finishing a locomotive? First must be twist drills, of which most folk have two sets; one set from $\frac{1}{16}$ in to $\frac{1}{2}$ in by increments of $\frac{1}{64}$ in and then a set of number drills. These range from No 1 at 0.225 in to No 60 at 0.039 in, by rather arbitrary increments of a few thou. They do go down to smaller sizes, but not in common use. These two sets will probably cost you £60 or more, so buy stands for each set and build up as you need each size. You can do the same with reamers. These look rather like twist drills, but have no point, and have 'slow' spirals along their length. These are used to finish a previously drilled hole to exact size with a good surface finish, on small bearings for instance. The standard two-flute twist drill actually cuts a triangular, or three-lobed hole, so the procedure is to drill a few thou undersize, then pass the appropriate size reamer through, using it just like a drill. Go for hand reamers which have a parallel shank to go in an ordinary chuck or collet and a squared end which will go in a tap wrench. Buy the common sizes, such as $\frac{1}{8}$ in, $\frac{1}{4}$ in and $\frac{1}{2}$ in then borrow the bigger ones if you can. Once again, the same applies to milling cutters and slitting saws, buy only what you need, when needed.

Slitting saws, incidentally, look like small circular saws and are used in the horizontal milling mode to make slots in valve-gear components and so forth. The most often used milling cutter is the end mill, which looks like a very short, four-flute twist-drill with cutting edges on the front end. It is used to form steps in components like axleboxes. It has a near relation, the slot-drill, which has only two flutes and can plunge straight down into the work, before being traversed to form slots which do not break out of metal at each end. You will need slot drills to form the steam ports in cylinder blocks.

Woodruff cutters look like small, thick slitting saws on the end of a short shaft, and these are often used to form the flutes in coupling and connecting rods on locomotives. Just one will probably cover most of the work on an engine. D-bits are handy little cutting devices for the amateur and occasionally the professional. Although of very simple shape, they can produce accurate holes with a flat bottom in such items as boiler fittings and axle-pumps. Again, you can buy a set if you wish, but the cheapest solution is to make them up as required from stock silver-steel. Silver-steel is a hardenable grade of carbon steel sold in 13 in lengths, accurately ground to size with a fine surface finish, hence the name. To form a D-bit, carefully file away exactly half the diameter for an inch or so, angle back the front face about 10°, then heat up to bright red and drop into clean water. The steel will then be glass-hard and consequently brittle so we need to temper it back down to a lower hardness. Clean up the flat area with emery paper, then heat the metal gently, watching the cleaned up area very carefully. As the business end colours up dark yellow, drop the bit into water again. Give the angled front face a rub on the oilstone and your D-bit will be ready for action. A variety of cutters, from simple slot-drills to taps and dies, can be made up using this material, so bear this in mind for the future.

Mention of heating raises another point; you will need some sort of torch for a variety of small jobs, such as tempering and silver-soldering. In the old days the paraffin blow-lamp was all mighty, but now, of course, the butane and propane torch is far cleaner and simpler to use, ideal for our purpose. When it comes to boiler making something larger will be needed and we shall discuss that in Chapter 8.

Evening classes, farming out, collaboration
Finally on the subject of workshops, one very rich lode for the beginner to mine will be the local technical college. Many of these run evening classes in model

A home-made D-bit in the tailstock chuck.

engineering and are equipped with large modern machine tools which will zip through the average model engineering job at a splendid pace. In addition, a skilled instructor will be on hand to help and advise whenever needed and no doubt quite a lot of material can be scrounged from the establishment, thus saving on costs. Don't be put off by the phrase 'evening classes', you will not be set to make ashtrays or toasting forks like school metal work classes. By and large you should be free to carry on and do whatever machining you wish. The cost of classes is quite nominal and the opportunities great, so go ahead and use them. A list of places offering evening classes in model engineering is usually published each year in an August issue of *Model Engineer*.

Another course of action for beginners is to farm out the big bits to a sub-contractor. Many of these advertise in the model press, or you can call round to your nearest friendly precision engineer. Not long ago I asked a local engineering shop to bore and face a pair of cast-iron cylinder blocks for a 5 in-gauge 0-8-0 freight engine. These were quite complex, being outside cylinder, inside steam-chest, but were done in an afternoon at a cost of £30, which is very reasonable. You may notice in various places throughout this book my emphasis on speeding the work, on taking short cuts and on making things as easy as possible, even if this means paying out cash to get pieces completed. The reason for this is simple and basic. There is a lot of work involved in building a model steam locomotive, as I have suggested, perhaps 1,000 hours. This is quite long enough for even the most enthusiastic man to become disheartened at the slow progress, and for him to take to other pursuits with faster rewards. Thus, I feel that in buying some components ready made, or sub-contracting some part, a whole locomotive may be saved from rusty oblivion to run in full glory on the track. Pursuing this particular line of thought leads to another suggestion, that of collaboration. Find another person thinking of building the same engine and work together. Since a large part of the building time is spent in setting up to machine perhaps just a pair of components, then it makes sense to use that set up to produce two sets. If you then divide up the work between you, there should be a considerable overall saving in time. Besides this, you will have someone to discuss all your problems with!

Chapter 5

The mechanical bit
Part 1: The rolling chassis

Frames, beams, stays and horns

The main frames are the basic structure on which a locomotive, large or small, is built, so naturally most people start out by making these. American locos usually had bar frames which present the modeller with something of a problem, since even in $3\frac{1}{2}$ in gauge, these will have to be cut from mild steel bar of something like two inch width and half inch thickness. Hacking out all the various recesses and holes in this stuff is fairly heavy going, so perhaps we should stick to a British prototype with plate frames. This will involve cutting material of $\frac{1}{8}$ in thickness generally for $3\frac{1}{2}$ in gauge and 5 in gauge, or $\frac{3}{16}$ in for $7\frac{1}{4}$ in gauge.

Take some care with choice of material since this will be the whole basis of the engine and some battered piece of second-hand ship plate will be likely to spoil the whole job. I personally prefer bright mild steel plate, but the black or blue varieties are fine. One snag here is that plate is usually sheared to size, giving a poor, bevelled edge, so buy a little oversize to give you a chance to clean up as necessary. Furthermore, shearing tends to impart a gentle curve, especially to long frame plates, which is a nuisance. If the bend is not severe, set the plates back to back to neutralize the bend; the buffer beams and stays will then haul them into shape. If the bend is more severe than, say, $\frac{1}{8}$ in per foot, try some careful straightening or get better material!

Some folk advocate the normalizing of rolled plate before cutting to shape. This involves heating the blanks to around 250°C for a few hours, followed by gentle cooling, in order to relieve the stresses set up by the rolling process. The plates then remain true after all the various cut-outs and recesses have been formed, which is not always the case with untreated mild steel plate. This is not a vital process, however.

The first job is carefully to file up one long edge straight and true, then use this as the reference line to mark out the rest of the frame from the drawings. This calls for meticulous use of the steel rule, scriber, square and centre punch, so sit yourself down somewhere comfortable for this job and check everything. Many people will find marking blue useful here. This is a sort of quick-drying alcohol-based dye which is painted on to the metal before marking out, so that the scribed lines show up more clearly. This is fine until you start cutting out the metal, after which the effects of the vice, even with soft jaws, cutting oil and filings make

rather a mess of the marking blue, so I prefer to do without.

Having marked out one frame, clamp the other blank to it and drill out a few of the marked holes, then bolt or rivet the two together. Now take your hack-saw, spit on your palms, and go to it. Saw carefully, just outside the scribed lines, in order to minimize the subsequent file work to clean up the edges. Use a drop of cutting oil on the hack-saw blade to extend its life and ease the work a little, and keep up a steady pace. The recesses for the hornblocks can be tackled by drilling a hole in the corners and then using a coping saw, or by drilling a row of holes along the line and breaking the waste out with a big pair of pliers or mole grips. Finish all the edges by draw-filing, then split the pair and de-burr; that is, remove all the sharp, burred-over bits to make the plates safe to handle, without actually rounding off the edges. Incidentally, beginners should note that there are lots of ways of making loco parts and if I describe some particular method it is not necessarily the only correct method.

There will be a number of countersunk holes for fixing the frame-stays and so forth, and the countersinks are formed after parting the frames. You can use a countersink bit in the drilling machine, or a twist-drill of the same diameter as the head of the screw to be used. In either case, use the depth stop so that all the holes are the same, and hold the frames firmly to avoid chatter. Avoid beginner's clanger number one, be sure to make a handed pair, that is one left-hand frame and one right-hand frame, when cutting the countersinks!

Now we have to machine and fit the hornblocks, these being the castings in which the axleboxes slide up and down to provide for the suspension of the

Left *A beautifully-made chassis for a 3¹/₂ in-gauge 2-6-4 tank loco which is up to exhibition standard. The design is 'Jubilee' by Martin Evans, but with much extra added detail.*

Right *This is me in my workshop, doing a spot of work on the new club loco.* (Caroline Coles.)

Below *A selection of castings for a 3¹/₂ in-gauge loco, machined on my Myford ML7 lathe. The castings were kindly supplied by A.J. Reeves and Co Ltd of Birmingham.*

locomotive. Small gauge hornblocks will probably be of the hot-moulded brass type, requiring little more than a clean up with a file to fit in the frames. Hornblocks of 5 in and larger will be in cast iron, requiring machining, first of all, on the face which fits to the frame. Hornblocks are an awkward shape to hold in a machine vice, so they are clamped to the milling table, or vertical slide, by a suitable block. This is just large enough to catch the flange of the hornblock which pokes through the loco frame, and has a hole in the centre to clamp the job down via a tee-nut and stud.

It is then a simple matter to run an end-mill over the bolting face of the hornblock, using your calliper gauge to dimension the flange to fit the frame closely. Next, make up a jig-plate out of frame material, like a section of loco frame, and make the hornblock to fit this. The jig can then be used both to drill the holes for the rivets in all the hornblocks and as a mounting plate for machining the inside face. Having used this jig to drill the hornblock, bolt the hornblock to it using small countersunk screws, then carefully file the flange flush to the jig surface. Now you can clamp the jig plate to the vertical slide and cut the inside face of the hornblock to the correct depth. Deal with all the hornblocks thus, then offer each up to its place in the main frames and drill through. Countersink the holes on the outside (remember — a handed pair!) and use soft iron rivets to fix the hornblocks for ever. Clip the rivet about a diameter above the frame face and use a round-headed or ball-pein hammer to drive it down into the countersink with the rivet head resting in a suitable 'dolly'. Clean up all flush by judicious use of a file, but avoid scoring over the frame when doing this little job as it looks nasty.

Now we need to deal with the jaws of the hornblocks where the axleboxes fit, and for this operation the frames are firmly bolted together again. If the frames are not deep you can just about get away with catching the pair in the machine vice on the vertical slide and then offering up each pair of hornblocks to a suitable end-mill, taking careful, fine cuts. For larger frames, however, seek out the use of a reasonably big vertical mill to do a really good job. You can get away with filing here, using a gauge block to get the dimension right, but already you will be understanding why I advocate equipping the workshop with a vertical mill. Clean up the bottoms of the hornblocks, where the hornstays go, while set up to do the jaws.

Some builders prefer a different sequence of operations here, making the axleboxes first in order that the hornblock jaws can be machined or filed to fit them. This is fine if you feel this is a better way round to achieving accuracy. It is a

Top right *Milling one of the sand-cast hornblocks on the vertical slide. Note the clamp block and note also the rag pushed into the top-slide hole to protect the cross-slide feed screw.*

Centre right *The frames, buffer beams, stretchers and motion brackets of an 0-6-0 tank engine.*

Right *Two types of hornblock casting. On the right is one of the hot-pressed variety which need little in the way of machining, on the left the sand-cast version. The latter is rather closer to scale practice. Both castings are for $3^1/_2$ in-gauge.*

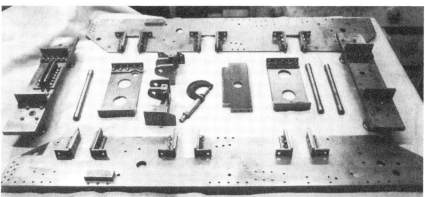

Left *Here the casting has been turned round and bolted to a jig plate in order to machine the inside face.*

Right *End-milling the frame stretcher casting to length.*

Below *Here are some more examples of sand castings for a 3$^1/_2$ in-gauge loco; a dummy tender spring, a main frame stretcher and two eccentric straps. Note how the eccentric straps are slightly oval to allow for cutting them into two halves.*

good plan to number stamp each axlebox and hornblock so that fitted pairs can be correctly reassembled at any time.

The next job is the buffer beams, or, in the case of a tender engine, one buffer beam, one drag beam. These may be flat slabs of metal held to the frames purely by brackets, or may be angle-iron, slotted to accept the frames. In the latter case, take care in marking and cutting the slot to get a good, accurate fit. The slot can be cut with hack-saw and file or machined out if you have the gear. The fixing brackets are probably best riveted to the beam and bolted to the frames, but individual designs may suggest otherwise. Be careful when drilling the frames for the brackets. Rest the assembly of frames and beams on its back on a true surface, such as the lathe bed, then clamp each bracket and spot drill through to the frame. Dismantle, drill right through, then bolt up firmly, again resting on a true surface to avoid any twist in the frames. A modern alternative to toolmaker's clamps in this, and many other operations, is the use of super-glue. This is powerful enough to hold metal parts for marking purposes, sets in seconds, and can be broken by a gentle tap or heating. The only snag is that it will glue your fingers together given a chance — ask any hospital casualty department!

After the buffer beams come the various frame-stays according to your chosen design. These may be fabricated or cast, and in the latter case the ends should be brought to length, true and square, by milling for preference. If your design has pony trucks or bogies, then you will need to turn their mounting pins on the lathe and fit them to their respective frame-stays.

Your next job is to make up the ponies or bogies if you have them. These are often made rather like miniature main frame assemblies, usually of the same materials, and form a good exercise ahead of the main structure. Their axles, axleboxes and wheels are produced in the same way as the main drivers, so let me go on to describe these as the next job.

Wheels, axleboxes, axles, eccentrics and pumps

Full size locomotive wheels were made of cast steel and in miniature loco work this is replaced by cast iron, usually produced by the classic sand moulding method, from a wooden pattern. You can, as mentioned in Chapter 2, have a go at this process if you wish and do your own cylinders and so forth at the same time, but it is a fairly lengthy business and the majority of people settle for buying a set of castings. Examine these before you start work to make sure there is sufficient metal to bring the wheels to dimension and that there is no 'flash' between the spokes. This is iron that has run along the join line in the sand mould and is often chilled, and hence very hard. It is a pig to remove so reject the wheels and ask for another set if they are flashed. Wheel turning tends to frighten beginners but, if tackled properly and reasonably carefully, it is quite simple. Cast iron used to be a problem in the days of carbon-steel lathe tools, because of the hard, chilled skin often found on the surface of the material. This meant clanking along endlessly with the lathe in back gear, taking a deep cut to 'get under the skin' and thus avoid losing the tool edge immediately. The advent of carbide-tipped tools has changed all that, with the result that you can turn a 4 in diameter loco wheel at middle speed with two cuts, one roughing and one finishing.

Buy a carbide-tipped right-hand knife tool; that is one that will do both long axis and facing work, with a tip radius of about $\frac{3}{32}$ in. This will be ideal for wheel work and many other operations besides. Do not worry about sharpening the thing; this is pretty difficult even with a silicone carbide or 'green grit' grindstone so when it gets tired, chuck it away and get another.

Right, now we are ready to tackle the wheels. First of all, clean off any adhering sand from the moulding process, using an old file or wire brush, then mount the first wheel in the three-jaw chuck, gripping on the tread, with the back facing the tailstock. Using the outside jaws of the chuck, you can turn a wheel up to 5 in in

Top right *The first stage of wheel turning. The casting is chucked by the tread and the back faced off, then the axle hole can be bored. In this view a reamer is being used to finish the axle hole to size.*

Left *A nice crisp cast-iron wheel, ready for machining.*

diameter, in a standard 5 in chuck, but if the wheel is too big, then you are going to have to devise something clever with the face-plate. Now you know why I say build a loco within the capacity of your machines.

Stand the wheel away from the chuck jaws enough to be able to turn the flange to diameter, and take care to see that it runs fairly true, without wobble. Use some packing behind the rims if you like, but remove it before you switch on, otherwise it will remove your eye before you can blink. Incidentally, the same goes for chuck keys. Now face off the back of the wheel, that is the rim and boss, removing enough metal to clean up these areas properly, but not so much that you make the wheel too thin to finish! If you return to your casting supplier for one wheel, he will guess exactly what has happened and his knowing smirk will cause you acute embarrassment! Your last cut should be fine and steady to produce a nice finish. Turning cast iron is generally pleasurable, but it does produce some rather penetrating black dust, so protect the lathe bed with a rag, taking care not to let it catch in the chuck, and protect your hands with barrier cream.

Next you need to make a true hole in the boss for the axle, which is done by centring, pilot drilling and then drilling to a few thou under size. You can then poke a reamer through or use a boring tool. The latter is aptly named since it takes more time than the former and it is harder to get all the wheels to the same dimension. Finally, bring the flange to diameter. Having done one wheel, repeat the process on all the rest. It is at this point that the beginner may regret his choice of a ten-coupled loco, however, you should be doing the job pretty smartly by the time you get to the last one.

The next job is to face the fronts, so turn the wheel round and chuck by the flange, pressing the turned back hard up against the chuck jaws so that the wheel runs true. It does not need to be centred for this operation, though. Face off the boss to bring it to the thickness shown on your drawings. You can make a little gauge with a step filed to the thickness of the boss and poke this down the axle hole, or take the wheel off, measure the boss with a calliper gauge and use the index collar of the top-slide to take off the required metal. Note the final reading on the dial and then do all the wheels to this setting.

Do the same sort of thing for the front of the rim, and you are left with just the tread to turn to diameter true with the bore and forming the flange to correct

shape. For this you need to make a simple jig, the major part of which is a round chunk of metal about ½ in or ¾ in thick and of a diameter a little less than that of the wheel tread. Grab this firmly in your chuck and face it off true. Then drill and tap its centre to take a steel pin slightly larger than the hole in the wheel. Screw the pin home hard and then skim it down until the wheel will push over it without shake.

Part or face the pin slightly shorter than the thickness of the boss, then drill and tap it to receive the biggest clamp-bolt it will accept. You can now hold each wheel absolutely true for turning the treads, just so long as you do not remove the jig from the chuck. If you must use the lathe for something else, then take the whole chuck off carefully. Be careful also as you turn the treads to diameter, because a heavy cut may spin the wheel on the jig. Bring the wheel and flange to the dimensions shown on your drawing, and write down the cross-slide dial reading when the tread is to dimension, or, if indexable, then rotate it to zero. Subsequent wheels can then be brought to the same dial reading, giving a set of wheels to exactly the same diameter. If chatter occurs, then slow down a speed. Remember that the radius at the root of the flange is most important.

The top of the flange should be roughly semi-circular and this, though best done with a form-tool, can be done by judicious use of a file. The chamfer on the front of the wheel rim can also be done with a file, but is best done as a subsequent operation using a tool set over to 45°. If the design calls for a groove to simulate the join between wheel and tyre found on the full size article, then this can be done at the same time.

Left *The second stage of wheel turning is to face off the front. Using a stout carbide-tipped lathe tool makes this operation a positive pleasure.*

Right *The final stage of wheel turning. Here the wheel is pushed on to a true-turned pin and held by a nut, so that the tread can be finished true with the axle hole. The flange is also machined to the correct profile at this time.*

Below *Figure 5.1: Chucking arrangements for the last stage of wheel turning, with the casting clamped true on a specially-made backplate.*

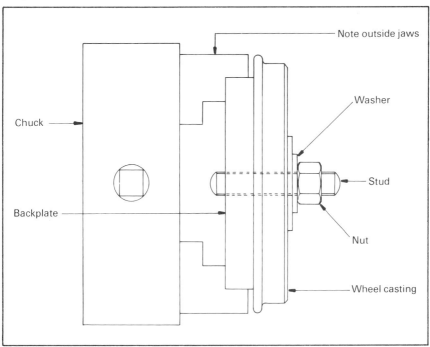

The final operation on the wheels is to drill for the crankpins, and once again a simple jig will lead to high accuracy. This is made from a suitable piece of metal, say ³⁄₈ in by ¹⁄₂ in or so, with two centre-dots accurately marking out to the crankpin throw. One is drilled to accept a pin which is a close sliding fit in the hole in the wheels, this pin being pressed or 'Loctited' into the jig. The second hole is drilled a few thou below the size of the crankpin hole. Make the jig long enough to be able to clamp it at the wheel rim, using packing if necessary between jig and rim. Scribe a line down the centre of the boss and centre the jig on it before clamping, then run a drill through the jig to make a deep countersink on the wheel boss. Repeat for each wheel, then drill clean through and follow up with a reamer ready for the crankpins themselves. Clever stuff, eh?

Talking of clever stuff, you can save yourself some bother later on by painting the wheels before machining them. It does mean that you have to choose your colour scheme at an early stage, but it can make life easier in the long run. We will talk more about painting in Chapter 10, but you may want to think about painting the frames at this stage also. Personally I prefer to strip down after the first run and paint before reassembly, but it may be a good idea to give the bare steelwork a coat of primer to prevent rusting. Items such as wheel rims and coupling rods should be left bare but well oiled.

At this stage, some people polish up their wheels with emery cloth to a bright finish, but this is rather inaccurate in scale terms. Better to confine yourself to cleaning up the lathe and workshop, which will now be coated in a fine black dust.

The next stage is the axleboxes. Now axleboxes can be a bit tricky, so go carefully here. There are a variety of designs, from the simple block of gunmetal

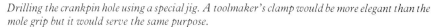

Drilling the crankpin hole using a special jig. A toolmaker's clamp would be more elegant than the mole grip but it would serve the same purpose.

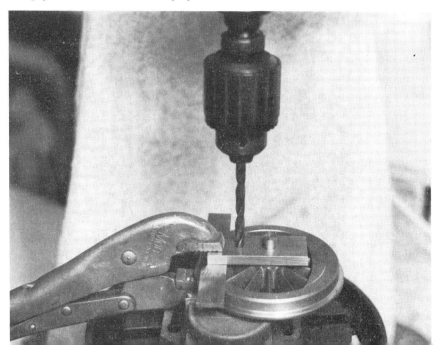

Above right *Detail of an axlebox, in this case one of the more sophisticated split variety. Note how the flanges are curved to allow each axlebox to rise and fall independently.*

Above left *End-milling a stick of axleboxes on the trusty vertical slide.*

Below *Milling an axlebox using the pin through the axle hole method. Note the aluminium washer to prevent damage to the turned faces of the axlebox.*

with a single flange fitted on 'Tich' to the split boxes with hinged hangers fitted to some 5 in- and $7\frac{1}{4}$ in-gauge locos. Some designs specify steel boxes fitted with gunmetal or bronze liners, others call for cast iron only, but let us assume that the prudent beginner has chosen something with basic, simple block axleboxes, probably of gunmetal. These will have to be machined on all faces, the grooves formed to fit the hornblocks and holes made for the axles, all to fairly high accuracy. There are lots of ways to tackle this job, and if you have a construction manual for your engine then it will probably give you some ideas as to how to proceed, but I rather like the method which uses a simple jig, thus:

Castings for axleboxes will probably be provided in sticks of four or six and first of all one of the major faces is machined either by milling or facing in the lathe. Next the axle holes are marked out reasonably true, drilled and reamed to size, and then the stick is sawn into the four or six blanks.

Set up an angle plate on your vertical slide or milling table and fit to this a pin which is a close sliding fit on the holes in the axleboxes. Provide for a bolt to grip the blank firmly on to the pin with the machined face against the angle plate. One face of the axlebox can then be machined, complete with groove, and the index dial setting, when the end-mill is at the bottom of the groove, carefully noted. When I say noted, I mean either zero the dial if it is indexable or write the reading down clearly if it is not. Now turn the blank through 180° using some true packing between the machined face and the slide to get the setting accurate. Then repeat the milling operation taking the bottom of the groove down to the same dial-setting as before, thus you have symmetrical faces.

The trick now is to leave the thing a little oversize, ideally by 10 thou or so. Measure the jaws of the relevant hornblock and then measure the width of the axlebox between the working faces, using your calliper gauge. Take half the difference in readings and machine this off the axlebox, then rotate through 180° again and machine off the other side by the same amount. Check with your calliper gauge by setting the inside jaws to the dimension of the hornblock jaws. The outside jaws should then just slip over the axlebox, in which case you have an axlebox which will fit the hornblock and has the axle hole truly central. Check the distance between the flanges by the same method, then machine to top and bottom faces on the same jig. Go gently because the blank may turn on the pin under the force of the milling operation. Use a sharp end mill, new if possible.

This then leaves only one major face of the box to be machined and this is done by pushing it on to a stub mandrel. This is just a length of bar turned to about half a thou over axle size, then slightly tapered with a smooth file so that the axlebox will push on with a stiff fit. The last face can then easily be faced off true.

Two or three final operations remain to complete the job. Firstly the flanges have to be bell-mouthed, so that each wheel can ride up independently without binding in the hornblock. This can be done by milling or filing carefully, but it is a pretty tedious job either way. Secondly, each axlebox is drilled and tapped for the spring hangers underneath or drilled for the spring pockets on the top if this is how the design is arranged. Finally, most axleboxes have a small hole drilled as an oilway from a reservoir on the top.

Above *What a colourful engine this 5 in-gauge 'Claude Hamilton' 4-4-0 is, and what a pleasant way to spend a September afternoon.*

Below *The builder of this lovely 3½ in-gauge 'B2' is a surgeon. Another fabulous mantlepiece exhibit and a fine track performer as well.*

Bottom *Model engineers' heaven. This is the delightfully chaotic scene inside Bishop Richardson's model loco works.*

Above *'Torquay Manor'
is a very popular 5 in-
gauge design by Martin
Evans and rightly so.
This one was competing in
the 1980 IMLEC
competition when
photographed.*

Left *'Rob-Roy' in detail,
showing a neatly cleaded
cylinder, rugged crosshead
and the ratchet drive to
the lubricator. This little
model is well on the way
to completion.*

Bottom right *Also
photographed at the 1980
IMLEC competition was
this delightful 2-6-4 tank
based on Martin Evans'
design 'Jubilee'. What an
attractive colour, I am a
great LMS fan.*

Above *Period piece — a sweet little working model of an early locomotive. Aren't those wheels splendid!*

Below *They don't come much bigger than this in 5 in-gauge. It took six men to lift this excellent French 4-8-2 out of the builder's car.*

Above *Little and large — a comparison between a 3½ in-gauge 'Rob-Roy' and a 7¼ in-gauge freelance 0-6-0 tank called* Cantab. *My petite wife can carry 'Rob-Roy' but it takes me and a neighbour to lift* Cantab.

Below *Exhibition standard is the only way to describe this 5 in-gauge '9F' by the well-known model loco man, Jim Ewins. Note the exquisite paintwork with superb dull finish, rather than the oft-seen toy-like gloss finish. That valve-gear is impeccable. What greater incentive could you need to get to work?*

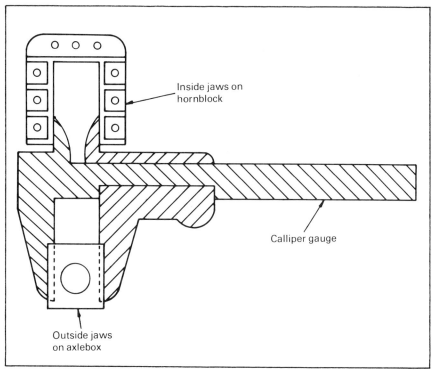

Inside jaws on
hornblock

Calliper gauge

Outside jaws
on axlebox

Figure 5.2: This shows how a calliper gauge can be used to match an axlebox with its hornblock. The calliper gauge is a tremendously versatile measuring tool and is vital to the model engineer.

Axles form a nice little exercise in simple turning if they are small enough to go down the mandrel of your lathe. In this case it is an easy matter to face the ends, centre and bring the wheel seats to the required diameter (see the section on wheel quartering). Care must be taken, though, to make the wheel seat concentric with the main bearing part of the axle, so an accurate three-jaw chuck or a collet is called for. Check the truth of the axle with a dial test indicator (DTI) before turning; anything more than a thou or two out means remedial action. Ground finish bright mild steel is the correct stuff for this job. If the axle is too big for the mandrel, then you can face and centre each end using the fixed steady and turn between centres, but take care with setting the axle true in the steady using a DTI or you will end up with wobbly wheels.

The driving axle of inside-cylindered locomotives is not quite such a simple matter, since cranks are required. By and large, the beginner would be well advised to avoid such a design, but nothing is impossible for the resourceful enthusiast. There are quite a number of ways of tackling crank-axles, the first being to turn from the solid. This is a job for a skilled turner, and I treat with some scepticism the stories of crank-axles turned from the solid, complete with four eccentrics for Stephenson's valve-gear. These days fabrication with Loctite probably provides the best solution, but if you must make a crank-axle, then you had best turn to some of the more advanced books.

Before putting the wheels on to their axles you will need to make up the crankpins and fit these. Silver steel is often recommended here because of its good surface finish, but it is tough stuff to tap for the leading pin, particularly in $3\frac{1}{2}$ in-gauge. Ground bright mild steel (BMS) is probably a better choice. As with axles, it is important that the running face of the pins is concentric with the part in the wheel, so again check your three-jaw or use a collet. Alternatively, turn both sections at the same chucking, but ensure a fine finish on the running face or your coupling rod bushes will last about five minutes.

In the past, crankpins were press-fitted by turning about half a thou oversize, slightly tapering at the end with a fine file and squeezing into the wheel using a vice. You can do this, or use the modern wonder-glue, Loctite. In this case turn half a thou undersize, that is, a sliding fit, clean the parts by scrubbing in alcohol, acetone or carbon tetrachloride and apply the high strength grade Loctite, which is a curious, thin, green liquid. Make sure all the surfaces are wetted by dropping the pin home and rotating, then wipe off the excess and leave to cure. Loctite is not so much a glue as a gap-filler. It cures hard in the absence of air (hence the use of the term 'anaerobic adhesive'), filling the minute gaps between the two pieces of metal and thus effectively creating a hundred per cent contact interference fit. It is quite amazingly strong but can be broken down by heating if necessary, but because of its gap-filling mode of action it can only be used on sliding-type fit applications, not for fixing sheets together. Loctite comes in various grades, the weaker ones being used for such jobs as locking nuts, where it is extremely effective but allows the nut to be undone again if necessary.

Before putting the wheels on to their axles, using Loctite again, you will probably need to turn up one eccentric at least for the axle pump, two more if the engine has slip-eccentric valve-gear, and four more if it has Stephenson's valve-gear. These are fairly simple to turn, but the trick is to make the offset hole for the axle true so that the eccentric does not wobble.

For eccentrics up to $1\frac{1}{2}$ in diameter or so, chuck a suitable piece of steel bar and turn the working face and flange, or flanges, using a sharp parting tool to get to the bottom of the groove. Go slowly and apply some cutting oil to avoid chatter problems. Part off a little oversize on thickness, turn round and face to the correct thickness. Now mark out the throw of the axle hole from the centre, as indicated by the turning marks, and make a centre-dot. You can then set this to run true in the four-jaw chuck, making sure that the eccentric is pressed well back against the chuck jaws as you tighten up. Simply centre, drill and ream, and there you are.

For larger eccentrics things are not quite as easy on a $3\frac{1}{2}$ in centre height lathe. Even on a 5 in-gauge loco the eccentrics may be $2\frac{1}{4}$ in diameter over the flanges and on a $7\frac{1}{4}$ in-gauge loco they will be over 3 in. Try chucking and parting a piece of $2\frac{1}{4}$ in steel bar and you will understand the problem. One solution to this problem is to use a casting but, surprisingly, eccentric sheave castings are not often listed by the suppliers. When faced with this job on a 5 in-gauge 0–8–0 loco I

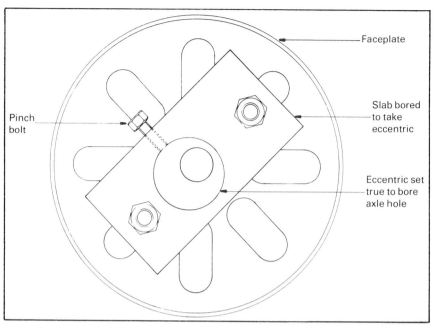

Figure 5.3: A simple jig for boring a set of eccentrics so that they all have the same throw and run truly on their axles.

made up a simple pattern and had six copies cast in iron at about £1 a time. This gave me four for the valve-gear, one for the axle pump and one spare. The casting was just a disc of the right dimensions with a chucking spigot on one side, and I went about machining the set as follows.

First, the sheave was chucked and the spigot trued up, using a carbide tool, of course. Then the casting was reversed, the main part faced, the flange diameter turned and then the working face turned to diameter, zeroing the indexable cross-slide dial at this setting. The other castings were then simply turned to this setting to give identical diameters, and the chucking pieces parted off. One big advantage of using cast iron here, is that it does not tend to chatter in the way that a large piece of steel will inevitably do on a small, less rigid lathe. Once chatter sets in, you may be reduced to pulling the mandrel round by hand using the counter shaft pulley, so that turning a set of eccentrics could take forever. Another simple jig had next to be made up, just a $\frac{1}{2}$ in slab of metal with a hole bored to accept the working face of the sheave and a pinch screw to hold it in. The boring was done with the slab bolted to the face-plate. This slab was rebolted to the faceplate, the first casting popped in, and what had been the back was faced to bring the flange to thickness. The casting was then removed and the hole for the axle marked out and pilot drilled $\frac{1}{8}$ in on the bench drill. It was then returned to the jig and the jig was shunted about on the faceplate until the pilot hole ran true, whereupon it was drilled and reamed to fit the axle with a gentle push fit. Each of the remaining eccentrics could then be treated the same way, giving exactly the same throw to each one, whilst all ran true on the axle and I was not driven mad by chatter problems.

Wheel quartering

Having made up axleboxes and eccentrics, we can now fit the wheels on to their axles, setting the crankpins at 90° on each pair of coupled wheels. Actually, it does not matter if the angle is out a degree or two either way, but what does matter is that all the coupled axles should have the same angle. If this is not the case, then the coupling rods will bind up badly and the loco will not run at all.

In times gone by wheels were fixed to axles by press fit or interference fit. The axle seat was turned one thou per inch of diameter oversize and then forced into the wheel, using a press of some sort. If the fit were slack, then the wheel might turn on the axle in service with nasty consequences. If the fit were tight, then the wheel casting could split in half. Remember that all the crankpins had to be at the same angle during this process, and there is a tendency for a wheel to revolve around the minute 'screw threads' of a turned axle seat, as it is pressed home.

Figure 5.4: Jig for wheel quartering. The axle is held in place by the threaded pins and the wheels rotated so that the crankpins rest on their respective flats. Elastic bands hold the assembly whilst the Loctite cures, preferably overnight.

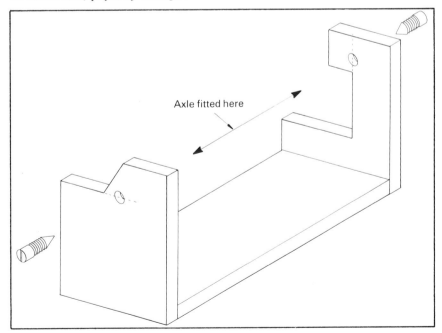

Axle fitted here

Various methods, using lathes and DTIs, jigs and so forth, have been suggested but all in all, traditional press-fit wheel quartering was an operation to make the beginner quail.

Happily, we can now forget all that with the advent of high strength Loctite, which makes the beginner's life delightfully easy. In this case the axles are turned one thou undersize for the wheel, that is, an easy push fit, then cleaned up as for the crankpins and fitted one wheel to each axle. It is very important that the entire area of the wheel/axle contact is wetted by the Loctite, so use plenty on each part and revolve the axle a little when it is home in the wheel. Be quick, because the stuff starts to go off in a few seconds. Wipe off the excess with a tissue moistened with meths and stand the assembly with the axle vertically upwards for 24 hours to be sure of complete curing. In the meantime you can make up yet another simple jig for accurate quartering. By now you may feel that you are spending more time making jigs than locomotive, but these devices are really the secret to success. They produce accurate results with great speed, so learn to love your jigs.

The sort of jig needed for wheel-quartering is shown in Figure 5:4 and can be easily made up from steel plate and bolted together. It consists of a simple frame to carry the axle, but the subtle part is the way in which the two end plates are cut away to meet the centre lines, one vertical, one horizontal. This means that whatever the diameter of the crankpins, if they are both held by an elastic band against their respective faces, they will be at 90° to one another. Thus, the same jig will do for different diameter pins and so can be used for any locomotive of the same gauge. It is worth making a nice job of this one, since you may well use it again in the future, or lend it to your less well-informed friends at the model club!

So, using this invention for final assembly, put the necessary axleboxes and eccentrics on to the first axle, clean up the axle seat and wheel hole with meths or what-have-you, and then Loctite the second wheel. Get the assembly on to the jig and in place fairly smartly or you may end up with the wheel half on and the Loctite well off. Move the axleboxes right away from the wheel being fixed to avoid permanently quartering these also. Allow the assembly to cure for 24 hours undisturbed and go on to the next axle. Do wipe up excess Loctite as it can be difficult to remove later, and sometimes causes unsightly stains.

One point here, you may wish to use the loco chassis minus wheels and axles in order to set out the centre distances for the coupling rods. In this case quartering of the wheels should be left until the coupling rods have been completed.

Once all the wheels are quartered you can proceed to the bits and bobs like springs, and get the wheels on to the chassis, which is the first milestone in loco construction. This is the time to fit pony trucks and bogies if your loco has them. As indicated before, they are often miniature versions of the main frame assemblies, and similar techniques are used in their manufacture. Thus the beginner might choose to make these first as a sort of trial exercise or 'hors d'oeuvre' of loco construction, then bolt them in place at this stage. The embryonic engine can now be rolled to and fro on its own wheels to the amusement of the children and probably anyone else who happens to be passing. Do not be disheartened by the inevitable, faintly patronizing praise from your

Above *This 7¹/₄ in-gauge 0-6-0 loco has the correct type of split axlebox combined with a fairly novel form of springing. Note the oil tube for an axlebox lubricating system.*

Below *Machining an axle pump body from a casting. Note how a chucking piece has been provided so that the body can be easily faced and bored. This chucking piece will subsequently be sawn off.*

Above *This eccentric strap is being faced and bored in the four-jaw chuck. The two halves are held together with screws during this operation.*

Below *Here a slot drill is being used to mill the eccentric strap to accept the eccentric rod. In fact an end-mill would have done the job equally well. The cutter is held in a collett which gives firm true chucking and a much better view of the job than a three-jaw chuck.*

The finished eccentric strap, complete with fixing studs and oil reservoir, compared to the casting.

wife, she is merely confirming her suspicion that you never actually stopped being a small boy!

By now you should be able to tackle the feedpump or pumps with no trouble. These are generally fairly simple machining jobs with the main body produced from a gunmetal casting which will likely be provided with a chucking spigot to make life easier. When tackling the job, pay attention to the fit of the ram in the bore and also to the ball seats, which should be carefully reamed in order to achieve watertight sealing against full boiler pressure. Check also that the balls cannot lift more than specified on the drawings or you will have trouble with oscillation. Axle pumps can be obtained for a number of the more popular designs, complete and ready to bolt on, if you so desire. They are not expensive and save a fair amount of work, but do do something about the pointless and shabby paint job that these items usually carry.

Castings will probably be available for the eccentric straps that drive the feedpumps from the eccentrics, and these are easy enough to machine. The first job is to drill through for the bolts holding the two halves together, taking care that the holes do not wander from the centre of the casting thickness. Then saw the thing in half and file or mill the cut faces true; the castings will be specially elongated to allow for the metal lost by this process. Bolt the pair back together and set up in four-jaw to face and bore the hole to run on the sheave. Use your calliper gauge to get a nice fit here. Reverse in the chuck and bring the strap to thickness, then arm yourself with your finest files and carve the outside to the required shape before riveting on the eccentric rod.

If the design has two pumps, then set the eccentrics 180° apart to give an even action as they work. Check that everything runs smoothly and then we can proceed to the next stage.

Coupling rods

I have to admit that I do not particularly like making these items. They need to be accurately made for the loco to run freely without nasty tight spots and this is not so easy, particularly where multiple-coupled axles are involved. Tons of metal have to be removed from the blank to produce the correct shape and the finish, to my mind, is very important in the area of the motion work, coupling and connecting rods. All this means that you spend quite a time on the coupling rods, but the loco does not appear to advance very much as a result of your efforts. Ah well, unless your engine is a single-wheeler, and lack of coupling rods seems to me a very good reason to choose a single-wheeler, then you must get stuck in to the job.

Let us assume first of all that we are working on a simple 0–4–0 design. Having laid hands on two suitable lengths of bright mild steel bar, the first thing to do is mark out the centres. Warning number one; do not try to use stainless steel, it is heart- and tool-breakingly tough and not worth the effort. Cleaned and well-oiled mild steel motion work will not rust. Warning number two; take care to produce a handed pair of coupling rods, also connecting rods and valve-gear. Now to marking-out and, perhaps, warning number three; you cannot simply use the designed distance between the axles and apply this to the blank with a ruler. Inaccuracies creep in here and there and the only sensible way is to take the dimension from the loco chassis itself. Furthermore, do not assume that both sides are the same, they probably will not be, so for each section of coupling rod take the centre distance from the relevant section of loco chassis, and do not muddle the pieces up!

The simplest method of obtaining the centres is to apply a large pair of dividers to the two axle ends with the loco laid on its side. It is not really possible to use the centre-drilled holes in the axle ends for this, so set your dividers from the joint between axle and wheel and transfer the dimensions to the coupling rod blank. Now centre-pop, drill and ream to crankpin size, then offer the rod up to the engine. With a four-coupled loco you will need to make up both drilled blanks, then you can turn the wheels and check for tight spots. If there is serious binding, then something has gone wrong and you had better try again. If there is just a hint of tightness do not worry, go ahead and drill to take the bearing bushes. If there is a definite tight spot, run the engine up and down on its wheels a few times, then remove the rod blanks and examine the holes. A bright patch should show where the rod is binding so that, with some careful work with a medium round-file, you can shift the hole over towards the bright spot, then drill for the bushes.

Another method for taking the axle centre dimensions and conveying them to the coupling rod blanks involves the use of a jig. This consists of two silver-steel pins with coned ends to match the centre-drilled holes in the ends of the axles, and a suitable bar of steel. You did remember to centre-drill the ends of all your axles,

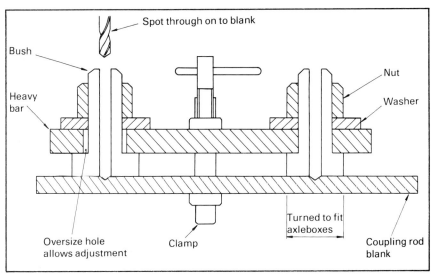

Spot through on to blank

Bush

Heavy
bar

Nut

Washer

Oversize hole
allows adjustment

Clamp

Turned to fit
axleboxes

Coupling rod
blank

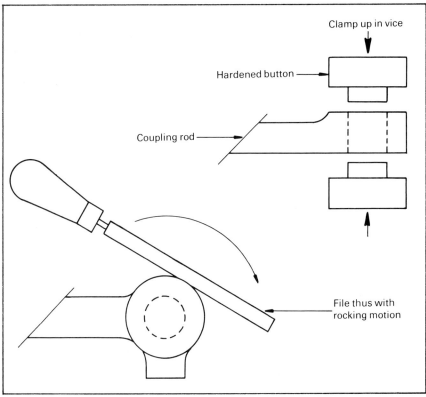

Clamp up in vice

Hardened button

Coupling rod

File thus with
rocking motion

didn't you? The first pin is hardened and tempered to blue and either screwed or pressed into the bar. The second is arranged either in a slot, or an oversized hole, so that it can be adjusted, and then clamped by a nut, preferably a winged nut.

Centre-pop one end of your rod blank, then offer the jig up to the locomotive. Incidentally, all the axleboxes should be at the same height in their hornblocks when doing this operation, so if they are not all pressed to the hornstays by their springs, pack them all up to the same height. Put the fixed pin in the centre of the first axle and then carefully set the movable one into the centre of the second axle and clamp. Now simply take the jig and put the movable pin into the centre-pop already in the rod blank, then tap the other pin with a light hammer just once. Deepen this light dot with your centre punch and, hey presto, the centres are accurately marked from the job. Drill and check the blank in the same way as before.

A third method involves a slightly more complex jig, but is, perhaps, the most accurate system. In this case instead of pins registering the axles' centres, we use buttons fitting snugly into the axlebox holes, hence the note in the previous section about leaving off the wheels until the coupling rods have been made. Again, one button is fixed on a suitable bar and the other is movable, to fit the actual job. They are both drilled, so that the jig can be set from the engine, then clamped to a blank and deep countersinks drilled at the correct centres.

Having drilled out the blanks we then need to carve them out to shape both in elevation and plan, and possibly cut a flute in the front face for realism. Coupling rods are an awkward shape and there are no simple methods here. The first plan is to mark out the shape of the rod as clearly as possible, using marking blue if you wish, then set to with hacksaw, files, coping saw and so forth, until the correct profile emerges. The tricky bit, of course, is the shaping of the bosses at each end, complete with oil boxes on top, and one good idea is to make up some filing buttons and work to these. They consist of pairs of silver steel discs of the same diameter as the required boss, each with a short step, registering the hole in the end of the coupling rod. The pair are hardened right out by heating to cherry red and quenching, then clamped either side of the boss with a toolmaker's clamp. It is then an easy matter to file down to the shape of the button, but being hardened, the file will go no further.

The main part of the rod also has to be thinned down front to back in most cases, and this can also be done by judicious filing. I have in the past brazed on pieces of steel in order to thicken the bosses of a pair of coupling rods. This worked fine, and the braze is only visible on close inspection, but on the whole I do not think any real work was saved.

A certain amount of machining can be done on coupling rods and certainly those with a fairly heavy milling machine can probably save quite a lot of donkey

Top left *Figure 5.5: Adjustable jig used to set out the holes in coupling rods from the axleboxes already fitted to the chassis. The diagram shows the jig in sectional view.*

Left *Figure 5.6: How filing buttons can be used to form the bosses on the ends of coupling rods, connecting rods and valve-gear components.*

work, though the bosses tend to remain a problem. If you also have a rotary table then a good part of the boss can be formed by working round an end mill, and the oil box formed by some clever juggling of the handwheels.

It is often suggested that rod bosses can be machined on the lathe by the following method. A short step is turned on a suitable piece of square bar to a tight push fit on the hole in the coupling rod. The bar is then clamped in the lathe tool-holder, at centre height, parallel with the lathe axis. The rod is roughly cut to shape by hand and then pushed on to the end of the bar. Now it can be offered up to an end mill running in three-jaw or a collet, and the boss worked carefully round against the cutter. Care is the significant word here since chatter and catch-up are both highly likely. Hold the rod end using a thick glove or rag, and advance very gently, always feeding against the rotation of the cutter. It strikes me that this method is fraught with danger for both man and machine, but it may be worth a try.

If your chosen loco has more than four coupled wheels, then the coupling rods will require a knuckle joint for each extra axle, to allow the wheels to rise and fall with the suspension movements as the loco rolls over the track. In some of the simpler six-coupled designs, such as 'Rob-Roy', this knuckle joint is combined with the driving crankpin boss; in most others the joint is placed immediately behind the driving crankpin.

If eight- or ten-coupled wheels are involved then there will be two or three knuckle joints respectively, again normally placed behind the crankpin bosses. Where the joint coincides with the crankpin marking-out of the blanks is carried out as described and the joint formed subsequently. Where the joint is separate, the thing to do is make the joint first, and then set out and drill the holes for the crankpins. In either case the machining is roughly the same, firstly the holes are drilled in two blanks, then the slot is formed by milling, using a slitting saw on an arbour, either in the lathe or on the mill. The matching part which has reduced thickness is produced with a spot-facing cutter or pin-drill. This is in effect an end-mill with a pilot pin in the end. This pin registers in the hole already drilled in the blank and the main body of the cutter can be lowered to produce a circular flat patch, concentric with the hole. This is usually done on both sides of the rod, cutting gently down to match the slotted part. This means that careful depthing will be required, so mill owners will be in clover and those using vertical slides on the lathe will need to read their leadscrew handwheels. Conceivably, one might use the bench drill by setting the depth stop with packing, but in any case, the blank should be well clamped to prevent rotation. The cutter itself may be a commercial item, or you can make the thing yourself from silver steel. Drill first for a press-fit on the pilot-pin, then file a screw-driver shape, and finally back off the two cutting edges. Harden and temper dark yellow, touch up on an oilstone and there you have it.

The rest of the joint will then have to be formed by filing as per the main crankpin bosses, then the two parts of the knuckle joint can be tried together. In some designs a bush is used here, but in most a plain steel pin, perhaps case-hardened, is used running directly in the reamed rods. One problem here is that

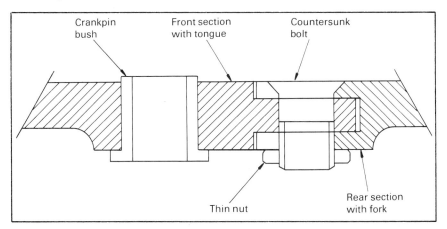

Figure 5.7: Top view section of a coupling rod to show the knuckle joint fitted just behind the crankpin. This joint allows for the movement of the coupled axles relative to one another as the loco rides over uneven track.

there is little clearance at the front of the coupling rods and none at the back for fixing the pin. The usual solution here is to make a countersunk head to the pin, so that it fits flush with the back of the rod. On the front a very thin nut is used to minimize protrusion.

Knuckle joints may remain plain, since there is very little rotational motion involved, but there is continuing rotation under load at the crankpins and hence these must be bushed with a suitable bearing material such as cast gunmetal or phosphor bronze. The latter is nasty, tough, cheesy material to machine but makes jolly good bearings. The bushes can be press-fitted to the coupling rod and then reamed to size after, or they can be taken to size on the lathe and Loctited in place, not forgetting to drill a small hole to take oil, from the pocket in the boss, through to the bearing.

A small problem arises when bringing the coupling rod bushes to size for the crankpins. If each bush were precisely reamed to fit its crankpin, and if each axlebox were a close fit in its hornblock, then the coupled wheels could not rise and fall independently with the inevitable irregularities of the track. In point of fact, the whole lot would bind up solid on the track. The answer is, of course, that there must be some free play in either the axleboxes, the coupling rod bushes, or a mixture of both. If you are the scientific type you can calculate the maximum freedom necessary to allow full motion of the axleboxes.

As an example, LBSC's 2-6-0 tender engine 'Princess Marina' has leading coupling rods 6 in long, and axlebox travel of $\frac{1}{2}$ in. If the leading axle were fully deflected upwards and the second axle dropped to the hornstay, then the change in crankpin centres is 0.020 in. In practice, however, a figure of about half this, 0.010 in, of free play will allow free running of the loco. You might, in theory, try to spread this by allowing, say, 4 thou at the axleboxes, 4 thou at the leading

Left *This 3¹/₂ in-gauge 2-8-0 loco is well on the way to completion by a beginner. The design is 'Euston' by Martin Evans, based on the Stanier class '8F' loco of the old LMS.*

Bottom left *Fine work here on the coupling and connecting rods of a 7¹/₄ in-gauge 0-6-0 loco. Note the castle nut on the knuckle joint and the brass oil cups.*

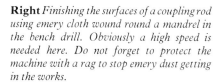

Right *Finishing the surfaces of a coupling rod using emery cloth wound round a mandrel in the bench drill. Obviously a high speed is needed here. Do not forget to protect the machine with a rag to stop emery dust getting in the works.*

crankpin and 2 thou at the driving crankpin, that is the one which carries the connecting rod. Again, in practice, you may well end up with this kind of free play without trying too hard, particularly as a beginner! However, you will need to open out crankpin bushes other than those on the driving axle to a diameter a few thou larger than the nominal reamed size. How is this done? One way is to drill oversize using a suitable letter drill. For instance on 'Rob-Roy' with crankpins of ¹/₄ in diameter, a letter F drill is recommended, giving a clearance of 0.007 in. Another idea is to bore the bushes on the lathe slightly oversize before Loctiting in place, and a third possibility is to go out and buy an oversize reamer from the toolshop, either imperial or a suitable metric size. You can, in fact, buy metric reamers in increments of 0.01 mm equivalent to 0.0004 in, less than half a thou, though these tend to be expensive.

The coupling rods of many full size locomotives were fluted for lightness and strength and, though many beginners' model designs leave this out, you may want to stick with full size practice. Those with horizontal mills can run the rod under a woodruff cutter or a side-and-face cutter. Mere mortals equipped only with a vertical slide for the lathe can clamp the rod to an angle plate and run under a woodruff cutter, or even simpler, clamp the rod at centre height into the toolholder and traverse across an end-mill. The disadvantage of the latter method is that the ends of the flutes are round, not square, which looks nothing like the forged original. Hopefully, the length of the flute will be less than the traverse of the lathe cross-slide, otherwise two bites will have to be taken, calling for some careful realignment. In many ways, the fluting operation is best carried out at an early stage, before drilling and shaping the rod, but after marking out. The blank is much stiffer at this stage and you will have material at either end to clamp on to

When it comes to final finishing, ignore the old chestnut about draw-filing, that is using the file sideways, not lengthways, rather like a spokeshave; this produces a rather scratchy finish. Use medium emery cloth, backed by a suitable block of wood and wetted with some machine oil. Be careful to keep the edges square but remove the immediate sharp edge. One way to clean up around the bosses is to wrap a strip of emery around a suitable rod in the chuck of the bench drill. Wind so that the flap trails and retain with an elastic band at the top. Run the drill at high speed and with the rod resting on parallel packing, work the boss around the rod to give a fine finish. Do not forget to clean the emery off the machine afterwards.

A final touch is to add little brass oil cups to each coupling rod boss, which looks nice if you have the time, then fit the rods to the loco chassis. The leading boss will probably be retained by a countersunk screw in a thick washer, in order to give clearance to the crosshead, driving bosses will be trapped by the connecting rod eventually, and trailing bosses will be retained by plain nuts or collars according to design. Hopefully, if all is well, the coupled wheels should spin freely with no trace of tight spots; however if they run a little stiffly in places, do not worry too much. Once the loco has run a few circuits of the track, things have a habit of easing up quite nicely! In the motoring world, this process is given the respectable title of 'running in'.

Thus we have reached the first stage of the locomotive, the rolling chassis. Next comes the interesting part, the bits that make the engine actually go, so, spit on the palms and get stuck in again.

Chapter 6

The mechanical bit
Part 2: The real works

The cylinder blocks

The cylinders lie very much at the heart of a locomotive. They are the site of conversion, where the heat conveyed from the fire by the steam is converted into mechanical work. Initially this work effort is linear, but is converted to rotary motion by the connecting rod, crosshead and so forth, only to revert to linear motion at the drawbar. Theoretically inefficient, but practically quite effective!

Cylinders come in a variety of shapes and sizes, and are fitted to the locomotive in a variety of ways. Just about every conceivable combination of cylinder and steam chest position was tried out in full size and most of these have appeared in model form. One major difference between models and full size practice however, is the frequent adoption of slide valves in models rather than the piston valves used on the real thing. Slide valves are rather easier to make and keep steam-tight than piston valves, particularly in smaller gauges, but they do take more work to drive them. There are a number of designs of model available with piston valves and indeed, if you are fairly handy with machine tools then by all means go ahead, but generally speaking the beginner is well advised to go for slide valves. These can be quite well mocked up to look like the piston valves of the original from the outside, and only the real experts will notice the reversed mounting of the combination lever for outside as opposed to inside admission. Lost the thread? Wait for the section on valve-gear.

The most common arrangement of cylinders is outside the frames with the steam chests on top, as in 'Tich', 'Simplex' or 'Marie E', giving easy access for assembly, valve-timing and maintenance. Next most common set up is with the cylinders outside the frames, but the steam chests inside, as found on designs such as 'Rob-Roy', 'Maisie' and 'Netta'. This arrangement is rather less convenient, particularly for valve timing, since it is difficult to see the slide valve and ports up between the frames.

Next come the various arrangements of inside cylinders with the steam chests either above, below or in between the cylinder bores, according to design. Mounting the cylinders in a large block between the frames does have the advantage of being a very stiff assembly, the block itself forming a massive frame stay whilst the piston thrusts act near to the centre line of the engine. This reduces the 'shouldering' effect seen on outside cylindered engines under load and hence reduces mechanical stress all round. Just watch an outside-cylindered locomotive from the front whilst it starts away with a load, either model or full size, and you will see what is meant by 'shouldering'. The disadvantage of inside cylinders is the very difficult assembly and valve timing, particularly where the steam chest is between the cylinders, combined with the requirement for a crank-axle to be made. These problems seem to be no great deterrent, however, and designs such as 'Titfield Thunderbolt' and 'Maid of Kent' by LBSC, 'Princess of Wales' by Martin Evans and the 'Derby Class 4F' by Don Young, are very popular. 'Yer pays yer money and yer takes yer choice' as the saying goes. If your choice is a three- or four-cylindered engine then you will have a combination of inside and outside cylinders, which will keep you busy for quite a long time. Three-cylinder designs like Martin Evans' 'Royal Engineer' are not for beginners, though LBSC's 'Hielan Lassie' and 'Roedean' are not impossible choices. When it comes to machining the cylinder assembly this will hopefully be described either in the construction manual for your chosen design or in the magazine articles where it first saw the light of day. Incidentally, many model clubs carry volumes of back numbers of the model magazines and it is usually possible to borrow those copies with the articles on your chosen design. This is extremely useful for the beginner who might otherwise be out on his own, so to speak.

There are really just two materials from which cylinders are made, either gunmetal or cast iron, usually supplied as a set of sand castings comprising cylinder blocks, steam chests and cylinder covers. Occasionally the slide valves, steam chest covers and pistons are also included. Gunmetal is usually the choice in

Typical cylinder castings, the cylinder block itself, the rear cover and steam chest. These examples are in gunmetal.

smaller gauges and has the advantage of being rust-free, but it is very expensive and sometimes not so easy to work. Cast iron is usually pleasant to work and is much cheaper, but since it can rust, there is always the risk that it will seize up in storage and be ruined. Plenty of oil is the answer here and I think with a little care, particularly in regard to storage of the engine, cast iron is the better choice for cylinders.

The first job to tackle is the cylinder blocks themselves and machining these is rather like skinning cats in that there are a number of ways to go about it. The method you choose will depend on the equipment you have, combined with your own personal preference, plus advice from the construction manual or magazine articles. Smaller castings can be swung directly in the four-jaw to face and bore. Slightly larger blocks are often mounted on an angle plate, itself fixed to the face plate. The biggest cylinders can be attacked by clamping to the lathe cross-slide, packed to the right height, and using a boring bar between centres. Those with milling machines should be able to face directly with an end-mill and bore using a boring head, or return to the lathe.

Let us assume that as a prudent beginner you have chosen a nice simple design with outside cylinders and slide valves on the top such as 'Marie E' or 'William'. Have a good inspection of your castings first of all. Look for any evidence of blow holes and if you spot any, ask your supplier for a replacement. Most suppliers will exchange castings without quibble if blow holes become evident during machining, but this is not much consolation if you have already spent a month's spare time on the casting. Next check that there is sufficient metal available to bring the casting to size. Again it is rather depressing to bore and face the ends of a block only to find that there is a cooling dimple on the port face, and it cannot be machined to dimension. I have already said in an earlier chapter that the quality of model locomotive castings is generally not impressive, however, we must accept what is available and take care accordingly. Fine quality investment castings (also known as lost wax castings) are just beginning to appear in the model world and we can only hope that the trend continues. Interestingly, other 'hi-tech' items are also beginning to appear such as laser-cut locomotive frames and other components. Perhaps we may soon see the application of vacuum-brazing to boiler construction (see Chapter 8), however, this speculation is taking us away from our cylinder blocks.

Having satisfied yourself that the casting has sufficient meat, whizz over the flat surfaces with a file to knock off the knobbly bits and any sharp 'flash' along the mould joint-line. Then make up a hardwood bung and hammer it into the core-hole, flush with one of the end faces, to give a surface on which you can mark the cylinder bore centre, or set an arbitrary bore and work back to the port and bolting face. The choice will probably depend on the casting itself.

Let us further assume that as a beginner you are limited to using a lathe for all machining purposes, so what is a good way to have a go at this cylinder block? Probably the best scheme is to use the faceplate with an angle plate bolted to it, the Keats V-type for preference, but a simple angle will do perfectly well. Fit the cylinder casting to the angle plate, resting either on the bolting face or port face and with the plugged end slightly overhanging the angle plate. Clamp it with a heavy bar over the top and a couple of long bolts. If the casting rocks on the angle plate, true it up with a file so that it sits firmly on the plate, or it may shift whilst being faced. Now comes the slightly tricky bit of shifting the casting and angle plate around until the marked bore centre runs true, and the core-hole is at right angles to the faceplate.

Once you are satisfied that all is well, tighten up all round, pull out the wooden plug and then fit some kind of balance weight opposite the angle plate or the lathe will shake like an earthquake when you switch on. It is then a simple matter to face across the end of the casting, but go easy, because of the intermittent cut. Face off half the amount needed to bring the block to length, the other half obviously comes off the other end. Aim for a nice fine finish using power cross feed if you have it, a steady hand if you haven't.

The next operation is to bore the cylinder, so fit the stiffest boring tool you can into the toolholder and go steady, using the lathe self-act. Take care as the boring

Figure 6.1: This shows the various dimensions which have to be marked out prior to machining the cylinder block.

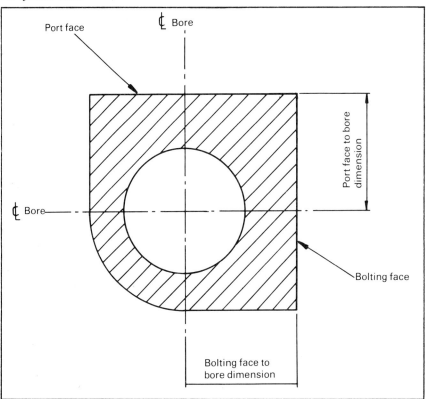

The first stage in machining a cylinder block is to face off one end, then cut the bore truly, as shown here on my Myford lathe. A heavy bar holds the casting to the angle plate and a couple of pieces of steel are used to balance the assembly. The boring bar is as stout as can be accommodated by the casting.

tool emerges from the far end of the casting or it will encounter the faceplate and cause untold damage. You did leave enough space between the casting and the faceplate to clear the boring tool, didn't you?

Opinions differ as to the best way of finishing cylinder bores. If your lathe cuts true, then continue boring until within a few thou of size, clean up the tool with a slip stone and continue to size, finally passing through two or three times without altering the cross-slide setting. By 'true' we mean that the lathe cuts parallel and in this case parallel probably means better than a thou in four inches. It might be a good idea to check the lathe by turning a suitable test piece before starting on the cylinders. If you discover that the machine is not all it should be, do not panic, but use the topslide to bore the cylinder. This is fine as long as the topslide travel is greater than the length of the cylinder blocks. Set the topslide true with the axis of the lathe by turning a test piece, checking with a micrometer or DTI and adjusting the rotation of the topslide accordingly.

The greater part of the cylinder boring operation can be done using the self-cut and saddle, the final few thou then being taken out with the trued up topslide. Again, make several passes at the last setting to take out the tool spring. LBSC favoured reaming as the finishing operation for cylinder bores and this, of course, took care of the parallelism problem, though it is really only practical for smaller cylinders, say up to about 1⅛ in bore. Beg, borrow or buy a reamer of the required dimension and bore the cylinder block until the slightly tapered lead of the reamer will just enter the hole. It is not possible to fit reamers this size into the tailstock, nor is it desirable since the tool should be partly self-guiding. Clamp some sort of carrier on the back end of the reamer and hold the countersink in the end of the reamer against a centre in the tailstock. Just enter the reamer into the bore and get someone else to switch on the lathe at low speed; unless you have three arms! Advance the reamer steadily, tailstock and all, through the bore without stopping until it is about to meet the faceplate, then stop the machine and back steadily out again. There probably will not be a big enough gap between the casting and the face-plate to clear the lead of the reamer so now remove the casting and hold it in a vice. Take care not to crush the newly formed bore because elliptical pistons are hard to make! Put the reamer through the rest of the way by hand but do not withdraw it, simply remove the carrier and take the reamer on through the casting. This way a nice parallel, smooth bore will result.

Now we have to face off the other end of the block and the recommended method for this is to push the block on to a stub mandrel. This is just a piece of bar held in the three-jaw, about the length of the cylinder, turned carefully to a thou or so oversize on the bore. Brass is nice for this job if you can lay hands on a big enough piece. The mandrel is then gently eased down with a smooth file until the

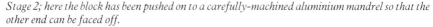

Stage 2; here the block has been pushed on to a carefully-machined aluminium mandrel so that the other end can be faced off.

cylinder block pushes firmly on to it, unmachined end outwards, then simply face across the end of the block and bring it to the correct length. Simple but accurate I think you will agree. It is best to tackle the two cylinder blocks in parallel, that is face and bore both, then face the other end of both and so on, to minimize setting up time. If you have reamed the cylinders, both should fit the same stub mandrel, if bored they may not, so deal with the larger one first, and bring down the mandrel as required for the second.

Next on the menu is to mark out on one end of the block the position of the port and bolting faces, at the correct distance from the bore. A job for odd-leg callipers and scriber, plus a little care, since the cylinders are rather important components. Then it is back to the angle plate set up to machine the two faces to the scribed lines. Set the block on its end this time, marked end upwards, with the bolting face looking at the tailstock. Use a big bolt through the bore and a suitable washer on top, not too large to obscure the marked out lines. Some soft aluminium packing at each end will prevent damage to those precious machined surfaces. Set the casting reasonably true with the bolting face parallel to the face-plate, tighten all round and balance as before. Clean up the bolting face to the scribed line, then slacken the clamping bolt and turn the casting through 90° to present the port face for treatment. You can get this absolutely true by registering the newly machined bolting face with one leg of your set-square and the faceplate with the other.

This is how the port face and bolting face are tackled, using the faithful angle plate. Note the balance weight again.

Tighten down and face off the port face as smooth as possible to the scribed line. Remove the casting and as LBSC would say, Robert is your father's brother, you now have a true machined cylinder block.

The following operation is another that may give beginners insomnia, but cutting the ports and passages is not particularly difficult. Start by marking the ports out carefully on the port face, using marking blue if you wish, to enhance the lines. When you are satisfied that all is well, fix up the block as shown in my photo, port face to the headstock, ports horizontal and set true by running the port face very gently against the face-plate. Those with vertical mills can set true in the machine vice with a square, but in any event, use some soft aluminium packing to avoid damaging the casting.

Put a slot-drill in the chuck or collet, a size suitable for excavating the central exhaust port and move the casting up close to it. Set the slot drill at one end of the exhaust port and note the reading on the cross-slide dial, then run the slot drill 10 thou or so into the metal. Traverse gently across to the other side of the marked-out port and note the dial reading again. Advance the cutter another 10 thou into the metal and traverse back to the original reading. Continue thus until the exhaust port is at the required depth, in fact, this does not take too long. In theory you should only have to look at your dials to complete the job, but if you are anything like human you will doubtless peer anxiously round the headstock from time to time to check that all is well; vertical millers will not develop a crick in the

Milling out the steam ports using the vertical slide. Note how the ports are clearly marked and hatched to avoid chopping out the wrong bit! The cutter has produced a mighty burr but this can easily be cleaned up later.

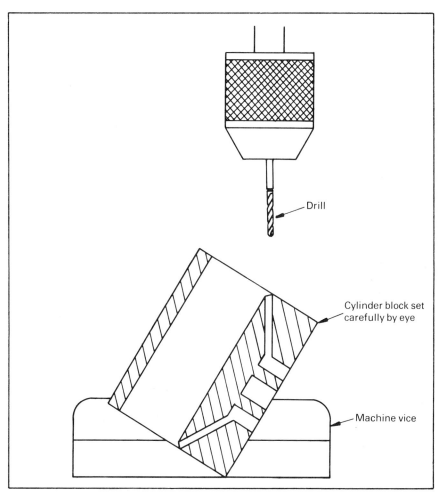

Figure 6.2: How to set up for drilling the steam passages from the cylinder through to the ports.

neck by these antics, since they can happily see the work progress with no problems.

Repeat the treatment using a suitable smaller cutter on the two steam ports, going fairly gently through the metal since these slot drills will be pretty small and consequently fragile. If you do break one it will probably damage your finances rather more than the casting. Next drill through from the bolting face to meet the exhaust port, this is usually quite straightforward work on the drilling machine. Some designs like 'Tich' call for a tapped main hole, with a second section canted to meet the port cavity. This can easily be set up by eye, using the machine vice on the drilling table, but in any event, no great precision is needed so long as the hole breaks into the exhaust port somewhere and not into the steam ports or port face. The same is true of the passageways. These are a bit more tricky, being smaller and deeper through the metal, but there is really no problem. As an untrained engineer I take the view that if I can do it, it must be easy! First of all, file a suitable chamfer, as shown on your drawings, on the lip of each end of the cylinder bores. Now make two or three equally spaced deep centre pops on this chamfer ready to drill, according to design.

Set up the cylinder block in the machine vice on the drilling table, as shown in the diagram, sighting carefully by eye down the drill and through the block to the steam port. Then drill carefully, backing out to clear the chippings frequently and going gently as the drill is about to break into the port. Take care not to continue into the exhaust port. If all is well and you are on target, move across to drill the second and, if necessary, third hole. Do not press hard when drilling or the casting may rotate in the shallow machine vice and break the drill, necessitating a trip to the spark-eroder for removal of the broken bits.

The last operation to complete the cylinder blocks is to drill and tap for the bolts and/or studs which hold down the end covers and steam chests. This is not done at this stage, however, since the covers and steam chests are drilled first and then used as jigs to drill the block accurately. If your design calls for cylinder drain cocks this is the time to drill and tap for these frilly bits. Drain cocks are not needed on slide valve cylinders since the valve can lift to clear condensation, unlike piston valves. You have not chosen a piston valve design, have you? Well, if you have, you can work out your own cylinder machining scheme by reference to some of the titles in the bibliography. If your locomotive is an outside cylinder, inside steam chest, such as 'Rob-Roy', then the machining is only slightly different to allow for the bolting flange which must be milled parallel with the bore. This can be tackled with the vertical slide and a suitable end-mill.

Cylinder covers, steam chest and pistons
The front cylinder covers are a pretty simple turning job. First, chuck by the rim in three-jaw and true up the chucking spigot provided, then reverse the casting to face and turn the register to a close fit in the cylinder bore. Use one of the blocks as a gauge and then fit the cover subsequently to that block. Bring to diameter and then part off the chucking piece or, if the cover is too close to the chuck, put in a vice and hack-saw off. How do we now face off the remaining rough side? The answer is in a split collet. This is made from bar, a little larger than the cover. Bore through this in the lathe a bit less than the diameter of the cover, then turn a recess which is a nice sliding fit on the cover, deep enough to hold the cover but leave the face proud for machining. Part off the bar about half an inch long, then saw through one side. Clean up the resulting burr, drop in the cover to be faced and clamp the collet in the three-jaw. Use the outside jaws and register the collet against the first step so that it runs true. The force of the chuck will compress the split collet enough to hold the cover whilst you face off to the correct thickness.

In order to drill the covers for the fixing screws we need a jig, nicely set out with

Above *This shows a split collet, made to hold cylinder covers for facing.*

Below *Here the collet is in use; the chuck jaws squeeze the collet so that it grips the cover firmly.*

the holes correctly spaced. Chuck another piece of the bar you used to make the split collet, face, drill and bore to fit the register on the cylinder covers, and then make a circular groove at the pitch circle diameter of the fixing screws. Use the tip of a pointed tool for this, then part off the jig an eighth of an inch or so thick. Mark out the positions of the holes using dividers, or whatever, then centre-pop well and drill the clearance size on the fixing screws you intend to use. De-burr the jig and place it over the register of the first cover, holding it with a small toolmaker's clamp. Then carefully spot through the jig to make deep countersinks on the cover beneath. Remove the jig and then drill right through the cover at each mark. To follow good toolmaking practice you could case harden the drilling jig, but since there are only four covers to deal with, unless you have a three-or four-cylindered engine on the stocks, it is not worth the bother.

Machining of the back covers starts out the same way as the front ones with facing and turning the registers, but it is more important with the rear covers that the register be a snug fit in the bore. Having achieved this the next operation is to drill and ream for the piston rod and again accuracy is called for, since this hole must be concentric with the bore, so take care when centring. Then it is back to your split collet to tackle the other side of the back covers; in terms of the whole engine, this must strictly be the back of the back cylinder covers. Confusing, isn't it?

Face the gland boss to the correct distance from the bolting face and that is about all you can do to clean up this side of the casting, because the gland boss will prevent you from facing up the rest. Open out the piston-rod hole to take the gland

Bottom left *A simple drilling jig for cylinder covers. This means that one lot of marking out will produce four identical covers.*

Right *The jig in use, fitted over the cylinder cover register.*

Below *Milling the lands on the rear cylinder cover to take the slide bars. I used a long-series end-mill to reach over the clamp bar and therefore had to take very fine cuts.*

next. Do not drill directly because the resulting bore is unlikely to be concentric with the cylinder bore. This would then cant the piston-rod over to one side with the result that it would bind up horribly. The answer here is to use a pin drill, either commercial or home-made, so that the pilot-pin keeps the counter bore concentric. Follow up with a taper tap, then a plug tap, so that the gland can be screwed well down into the gland boss. The glands themselves are relatively simple items, best made from phosphor bronze, but again, take care that the thread is concentric with the piston-rod hole. If not, the rod will again be canted over and bind. Concentricity in this whole assembly is important in order to achieve minimum friction with maximum steam-tightness, however, we are not talking about tolerances of microns, just good general machining practice. Use a tailstock die-holder to keep the thread true and open up the die to give a tightish fit in the boss so that the gland does not unscrew itself in service. If this happens, the crosshead will strike the gland a mighty blow and generally make mincemeat of your handiwork.

The rear cylinder covers also serve to carry the slide bar or slide bars, plural, as the case may be. Once again, some care is called for since slide bars need to be parallel with the piston rod, and set at the correct distance from it. The majority of locomotives, both miniature and full size, sport two bars with a so-called 'alligator' crosshead running between them. This means that we need to machine two flats on the gland boss, equidistant from the piston rod axis and parallel to it. In order to mark the position of these flats, make up a plug rather like the glands, but with a light centre instead of a hole, and screw this into the gland boss. Set your dividers at the correct axis-to-slide bar distance, put one point in the centre-dot and scribe two firm arcs on the face of the boss. Clamp the casting to your vertical slide with a suitable bolt through the piston-rod hole, then use an end-mill to clean up the slide bar lands, just down to meet the two scribed arcs.

Take care when drilling and tapping the gland boss for the screws which retain the slide bars, not to break into the gland cavity. One way to hold the cover whilst doing this operation is to screw a piece of threaded bar firmly into the gland boss, and this can also be used whilst filing up the unmarked surfaces of the cover. When it comes to drilling for the fixing screws, use the jig made for the front covers, but be careful to orientate it in relation to the gland boss, or you may find that a fixing screw encounters the steam passage.

Having drilled all the covers, you can then drill and tap the cylinder block to take them. The front cover can be orientated by eye, then a drill used to spot through on to the block. The rear cover, however, must be accurately set into the slide bar lands at right angles to the bolting face, and this is easily achieved with a set-square on a true surface.

When it comes to drilling and tapping the cylinder block it is important to tap the holes true, particularly those for the steam chest retaining studs. There is quite a forest of these, even in a small engine like 'Tich', and if any are out of kilter, it is the devil's own job to get the steam chest down over them. So lay hands on a tapping and staking tool, such as that described by George Thomas in the *Model Engineer* some years ago. Failing this, I manage to use my drilling machine for

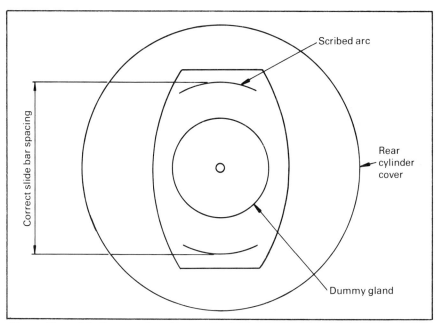

Above *Figure 6.3: One way of marking out the slide bar lands on the rear cylinder cover, ready for machining.*

Below *Figure 6.4: How to set the slide bar lands truly square to the loco main frames.*

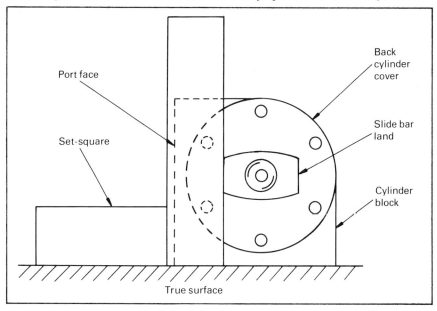

tapping, by hanging a weight from the handle to counteract the return spring. I then remove the drive belt and fit the tap in the chuck, so that it is possible to hold the chuck and draw it down to the workpiece, rotating it between the fingers. This is a bit time consuming, but it does start the holes true enough to bottom them out by hand. Next job, steam chests.

If your design tries to emulate the piston valve-chests of the original then the steam chest will no doubt sport two large cast-in bosses. If the engine is humble enough to admit to slide valves, then it may only have a small boss at one end, or indeed this may be a separate item screwed in after machining. If the casting is without bosses then it is a fairly easy job to grab in the four-jaw and square up, followed by drilling for separate bosses. Where the bosses are cast-on, catch the back one in the three-jaw and set the other to run as truly as possible, then centre it lightly. Run the tailstock centre up to this and then turn the boss and the end face of the steam chest to size. Now reverse the casting and repeat, but having turned the boss, drill and ream for the valve rod, then counter bore and thread as for the piston-rod. The long sides of the chest can be faced in four-jaw, but if the bosses are larger in diameter than the thickness of the chest, then the bolting faces will have to be milled in the vertical slide. Moderately tedious, but you must be pretty good at milling by now! The inside faces can be dealt with by filing since no dimensional accuracy is called for. The steam chest covers may be castings, complete with chucking piece, or just a piece of plain sheet brass, but in either case, mark out carefully for the fixing studs, then use the cover as a jig for drilling the steam chest, and the steam chest as a jig for the cylinder block. Again, take care

with tapping the holes for these studs upright and you will have no problems on assembly. You could, of course, use screws if you wish here, but even so the holes need to be true. Brass screws will shear off in tilted threads and stainless ones will strip out the thread. Both rather nasty happenings.

The next job is to make up the piston and piston-rod assemblies. The piston may be a casting supplied with the cylinder set, if not, use bronze or gunmetal in non-ferrous cylinders and cast iron in ferrous ones. The piston-rod should be made from ground stainless steel and perfectly straight, so be careful with this material. Thread one end as specified in your drawings then turn the piston blank a fraction oversize and tap part way, drilling for a press-fit on the piston-rod for the other part. The piston-rod can then be held in the tailstock and drawn tightly into the piston blank. The drawings should specify the dimensions involved to get the correct sort of part press, part screwed fit. The piston-rod can then be held whilst the piston is finished off, but concentricity is again important on this job. Ideally one should be able to twist the piston-rod around in the assembly cylinder and feel no resistance at any position down the bore. It does not take much error in the piston-rod, gland and so forth, to generate a tight spot in the works, so go carefully. A collet is the best bet, but if your three-jaw chuck is good, use that. Turn the piston to a sliding but shake-free fit in the bore, using each cylinder as a gauge for its own piston, and do not muddle up the bits when you have finished. In fact, it is worth stamping most of the parts of the engine as you go along, so that they can be put back in their rightful places whenever dismantling is called for. Many parts should, in theory, be interchangeable but, in practice, this is hardly

Left *This operation calls for some very careful setting up in the four-jaw to get everything true, but once chucked you can face the end of the steam chest, turn the gland, drill and ream for the valve rod and pin drill the gland, as seen here. The pin drill in use is home-made from silver steel.*

Right *Starting work on the steam chest. Facing up the bolting surfaces using the four-jaw chuck.*

likely to be the case. A set of number stamps and an 'L' and an 'R' can be a worthwhile investment.

The last components of the cylinder group are the slide valve, valve rod and valve nut. The slide valve may be a casting or it can be made from bar material, as for the piston. The outsides can be faced in the four-jaw whilst the exhaust cavity can be milled on the vertical slide with a small cutter. The recess for the valve rod need only be roughed out but the slot for the nut should be milled a good fit or there will be 'lost motion' to the detriment of the valve timing. The nut is just a simple piece of bar, and the valve rod a piece of stainless steel, as per the piston-rod, threaded to allow for the adjustment of the valve.

If all is well, then the cylinders can now be assembled, although the steam chest cover cannot be permanently fixed until valve timing is complete. Start by dealing with the pistons. LBSC's system here was to pack with graphited yarn and this seems to work perfectly well, although it sounds a little crude these days. The only problem is to compress the yarn whilst entering the piston in the bore, but with a suitable soft tool it can be gently persuaded to co-operate. The packing does not need to be greatly compressed, just pushed down a little to form a steam-tight seal without much mechanical friction.

The same goes for the glands which are also packed with graphited yarn and the gland nuts screwed down on to this, just finger tight. An excellent modern alternative to yarn is to use Viton (PTFE) 0-rings, these are advertised in the model press for this use. They are claimed to be steam-tight, even against super-heated steam, and relatively friction free, thus they seem like an engineered solution to the problem. You will need to turn the groove in the piston to fit the 0-ring according to the manufacturer's specifications for tolerance which allows the ring to 'roll', so beware here. Those with cast iron cylinders and pistons can use proper piston rings either home made, if you are a real enthusiast and have read the articles in the *Model Engineer* describing their production and heat treatment, or bought commercially.

Gaskets can be made to fit the various bolting faces from commercial materials such as 'Hallite' or, for smaller engines, oiled paper works fine, but again there is a much better modern alternative in the form of silicone-rubber type compounds, supplied in tubes like toothpaste. This can be bought from car accessory shops and is very easy to use. The only problem with using this material is to put on sufficient to cover the faces, but not so much as to extrude into the cylinder or steam-chest and cause a blockage.

As mentioned before, stainless steel studs are preferred for fitting steam chests, though screws are quite permissible, since it is usually not possible to see the top of the steam chest on the finished engine. Studs are also correct practice for holding cylinder covers, but this may be a bit fiddly on the model so use hexagonal head bolts (cheese-head or round-head screws look right out of place here so do not spoil the ship for a ha'p'orth of screws). When it comes to fixing the cylinders to the frames then abandon scale practice and use stainless steel allen screws. These are good and strong and it is possible to get a key on to them between the frames, something impossible with slotted screws and awkward with hexagonal heads. It has always seemed to me very worthwhile using the right screws and bolts on an engine. Quite often one sees a reasonably well made locomotive spoiled by odd size or shape fixings, despite the fact that screws of all types are freely available through the various suppliers at no vast expense. Do not pretend that you will change those grubby cheese-heads from the scrapbox with nice hexagonal heads later on, because you probably won't. Get the correct screws in at the time of building.

Slide bars, motion plates, crossheads and connecting rods

Slide bars are fairly simple items, in most designs for beginners at any rate. Usually they will consist of a pair of plain bars, anchored at one end to the gland boss on the rear cylinder cover and at the other to the motion plate. Silver steel is sometimes suggested here but mild steel is perfectly adequate and will last for ages if well oiled. It is a good plan to use allen screws to hold the bars to the cylinder cover, as with the cylinder block, since the screws can be properly reached with an allen key.

The motion plate may be more or less complex depending on whether it carries the expansion link of the valve-gear or not. A casting may well be available for this item in which case some milling and filing will be called for, but you may be as well off using fabrication. The parts can be cut from mild steel plate and/or angle and then riveted, screwed or brazed together according to design and choice. The principal call for accuracy here is in the fitting of the slide bars which must remain parallel to the line of the piston-rod. If the bracket carries the expansion link then the trunnion bearings should be in line and at right angles to the locomotive frames. If you have come this far then you can probably devise ways to sort this kind of component out using the machines at your disposal.

The motion plate on inside cylinder designs takes the form of a frame stretcher with accommodation for the slide bars in some form. It can range from a simple plate design to a horribly complex casting which carries some of the valve-gear components as well. A good example of this is the Great Western 'King' Class, the construction of which in scale form requires considerable skill or a signed pact with the devil, preferably both! However, the astute newcomer will no doubt have chosen a less ambitious project and retained the ownership of his soul.

Crossheads are another of those components that I don't particularly like because they are a difficult shape very often and call for a fair degree of accuracy in manufacture. Worse still, some pretty high forces are involved when the crosshead is doing its job so flimsy or doubtful structures are out. It has always struck me that the crosshead is probably the weakest point in the drive train, between piston and drawbar, both in model and full size, and yet failures here seem rare. The only fault I have come across has been the detachment of the piston-rod from the crosshead and even then little damage occurred on this occasion. All this would indicate that current designs must be about right.

As regards material, I have seen castings offered for crossheads, both in gunmetal and nickel-silver. The latter was nicely cast and included the drop-arm for the Walschaert's valve-gear, but on the whole I prefer to start with a nice true, bright block of mild steel and work from there. A gunmetal casting sounds potentially rather weak in this application and, furthermore, does not resemble the cast or welded steel of the real thing. Perhaps we may see some fine steel investment castings for model crossheads at some time in the future.

Start out with a mild steel blank that will do for both crossheads and bring to size in the four-jaw if necessary. The grooves for the slide bars can then be end-milled in the vertical slide, but set up very carefully to get the groove central in the block and parallel to its edges. Machine very carefully to the specified depth, using the lead screw handwheel to index the depth. Those with milling machines should be in clover here. If your engine sports a single slide bar, like 'Tich', you can proceed to the next stage, if not, then the block must be reversed and a second groove cut. Again, take care with setting up, and turn the blank end for end, not side for side, so that even if the grooves are not central, they are at least in line. Check that the slide bars fit the grooves before removing the blank from the machine vice.

Some designs avoid the need for this milling operation by using fabrication and indeed there is a lot to be said for this approach. The slide-bar grooves in this case are formed by bolting thin plates either side of the thick central core. Furthermore, the central recess for the small end of the connecting rod can be sawn out of the core, rather than formed by pin-drilling as required in the solid pattern crosshead. The piston-rod boss will probably have to be brazed onto fabricated crossheads and this is a source of potential weakness, but it is probably easier than forming it from the solid blank.

On solid crossheads the piston-rod boss can be turned by holding the blank in four-jaw and using a very small boring tool, with a little care. The hole for the rod can, of course, be drilled and reamed at the same time, leading on to the problem of how to fix the piston rod into the crosshead. One method which is fine in smaller guages is simply to screw the rod into the crosshead, which has the advantage of allowing adjustment of the length of the piston-rod. Alternatively a more scale-like method is to use a close, plain fit and secure with one or two pins. These can be taper pins or gentle press-fit parallel pins, but should be of silver steel to withstand the high shear forces involved. If employing taper pins, you will have to buy or borrow a matching taper-pin reamer, so bear this in mind. Do not drill for the pins until you have done a trial assembly and checked that all is well in the reciprocating department, and that the piston stops short of the cylinder covers by an equal amount at each end. Mark the correct position of the piston-rod in the crosshead boss then disassemble and drill the rod and boss as a unit. A tightish push fit makes life easier here since the assembly will then stay where you set it whilst cross-drilling.

To make a trial assembly you will need the connecting rods and these are very similar to coupling rods in many ways. The only slightly naughty bit here is that

Top right *More nice work on the crosshead of the 7¹/₄ in-gauge 0-6-0 engine seen earlier on.*

Right *The crosshead on 'Rob-Roy' also drives the water pump as seen here. Note the use of stainless allen screws to retain the slide bars.*

Not recommended for beginners, but a fine engine indeed. This is a 3¹/₂ in-gauge Great Western 'King' under construction by a professional model loco builder. This engine has four cylinders, two inside the frames and two outside. Those slide bars look like hard work and so do all those rivets!

many designs will call for a flute here to simulate the forged full-size rod. As with coupling rods, you can end-mill if you like, but the rounded end of the flute is wrong, so go for a woodruff cutter for preference. This produces a run-out at each end, very like the real thing and can be done on the ever-useful vertical slide. Some designs call for a tapered flute, wider at the big end which means two cuts will be involved. If the flute is longer than the travel of your cross-slide, take care in setting up for the second bite as any misalignment will show up like a sore thumb.

Those tackling inside-cylinder designs will also have to make big ends which can be dismantled for fitting to the crank-axle, like the connecting rods of a car engine. These can vary from a simple split bush to a complex gib and cotter assembly to take up wear as in full size. No real problem here, but it is just that bit more work than outside-cylinder designs in general.

Chapter 7

The mechanical bit
Part 3: More works

Introduction to valve-gears — slip eccentric gear

This is another set of parts which can tend to frighten the beginner, but in truth there is nothing particularly difficult about making up and assembling valve-gear. If your coupling rods turned out reasonably well, then you will have no trouble with valve-gear, it is just a fair chunk of work to deal with. The only slightly tricky component is the expansion link; other than this just go for decent fits and finishes and all will be well.

There are a whole host of different valve-gear designs, mostly named after their supposed inventors and each with their own particular advantages and pitfalls. Indeed, many people find the study of valve-gear quite fascinating and you will find quite a lot written on the subject in various more advanced books on model locomotives (see Chapter 11). It is also still a favourite topic in the model magazines and arguments rage continually over the niceties of the errors inherent in this design or that. In fact, all designs are some kind of compromise, both in theory and practice, so that to a large extent this kind of detail is irrelevant. If you fancy designing your own valve-gear it can be quite an absorbing occupation and ultimately rewarding if all goes well, but by and large, the beginner can ignore the design aspects and simply make the parts to drawing. Even so, it is as well to know what the valve-gear is trying to achieve, so I will give some simple explanation.

Broadly speaking, the newcomer will encounter three types of valve-gear, slip eccentric (sometimes called loose eccentric), Stephenson's and Walschaert's. Other reasonably common gears are Baker, Joy and Hackworth, but I suggest you ignore these for the present. Slip eccentric gear is the simplest of all gears and does not allow reversing from the cab, hence it was very rarely used in full size and consequently is only used in the simpler model locos such as 'Tich'. I am slightly mystified by the rarity of slip eccentric gear in model locos since it works perfectly well and saves a vast amount of work. The only disadvantage really, is that to reverse, one has to manually push the engine half a wheel revolution backwards. Since the valve-gear on very many loco designs is hidden between the frames, why not fit slip eccentrics and be done with it? You can always fit a dummy reverser in the cab if you wish, and I do not believe that the fixed cut-off given by this gear is any practical disadvantage; however, this kind of talk is rank heresy amongst loco fans, so be warned!

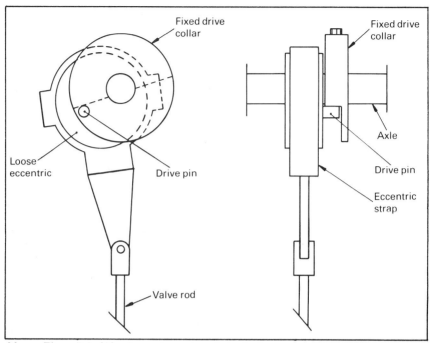

Above *Figure 7.1: Side view and top view of simple slip eccentric (or loose eccentric) valve-gear.*

Below *This is the underside of a large-boilered 'Tich' with slip eccentic valve-gear. Note the axle pump and lubricator with their various pipe connections, and the whistle at the top. This is the work of a total novice.*

A splendid example of slip eccentrics in full size practice was on one of the infamous compound designs of Webb in the latter part of the nineteenth century. This had two outside high pressure cylinders driving the two rear wheels, and was worked by Joy valve-gear and a large, single, low pressure cylinder between the frames driving the front axle, worked by slip eccentric. The two driving axles were not, for some strange reason, coupled and thus on trying to reverse the engine, if the rear wheels slipped, the enginemen would be faced with the situation of the two sets of driving wheels pounding round uselessly in opposite directions! Nice one, Mr Webb.

How does this gear work? Essentially the valve is driven by a simple, single eccentric on the driving axle in the same way that the old stationary mill engines worked. Instead of fixing the eccentric to the axle however, which would give only one direction of rotation, it is free to rotate. This loose eccentric carries a short pin which engages a stepped collar which is fixed to the axle. In forward motion the pin is pushed around by one side of the step and to reverse, the engine is pushed half a revolution backwards, so that the pin is then driven by the other side of the step. The engine will then run backwards. Have a look at the diagram which illustrates this. The step is not cut exactly on the diameter of the driving collar but some way below. This is to allow for the lap of the slide valve in both directions.

Stephenson's valve-gear

Stephenson's valve-gear is rather more complex, but is probably the next simplest gear to understand and to manufacture. This was common on British locomotives, particularly shunting engines, and so is found on plenty of models, 'Rob-Roy' being a prime example. In this case two eccentrics are fitted to the driving axle for each cylinder, one set up for running forwards and one for running backwards. One eccentric rod is connected to the top of a curved link, called the expansion link, and the other to the bottom. A little block called the die-block sits in the curved slot in the expansion link and this is connected to the slide valve. The expansion link can be raised and lowered through a system of levers from the cab reverser stand, with the effect that the motion from either one of the eccentrics can be transferred to the slide valve, or a blend of the motion from both.

This combining of motion from the forward and reverse eccentrics has some subtle effects on valve events, as mentioned briefly in Chapter 3. In full forward gear the forward eccentric rod is in direct line with the die-block, and thus the motion of the valve is derived solely from the forward eccentric. This will probably be arranged to give a cut-off of around 75 per cent, that is, if you remember, the cylinder will be closed off from live steam when the piston is 75 per cent of the way down its stroke. This is the configuration used to start a train moving, since it gives maximum torque at the wheels, but is wasteful of precious steam. Once the train is under way, the valve-gear can be 'notched up', that is, the cab reverser is brought some way back towards the central position, causing the expansion link to lift a little way. This means that the forward eccentric rod is no longer in line with the die-block, and that part of the motion of the valve is now derived from the backward eccentric. The effect on the slide valve is to make it close off the cylinder from live steam earlier in the piston stroke, that is, the cut-off is shortened. Thus the cut-off can be progressively reduced from the cab as the train gathers way until, at speed, a cut-off of perhaps 30 per cent can be used. This gives good steam economy and smooth, fast running.

If the cab reverser is brought fully rearwards, then the same sequence of events can occur, but with the engine now running backwards. As with slip eccentric gear, the two eccentrics are not set at 180° to one another because of the need to allow for lap. Very often the die-block is connected to the valve rod by an odd shaped link, suspended from a rocking lever, in order to negotiate the leading axle which gets in the way rather!

What are the problems in manufacture of this sort of valve-gear? Generally, not many; it is just a question of making up the parts with decent accuracy, then assembling and timing the gear. Eccentrics will have to be fitted back at the chassis assembly stage, as described in Chapter 5, and suitable straps and rods made to fit, as for the water pump. The expansion links may look tricky but really just call for some careful fitting. The thing to do here is to cut the curved slot first and make the little die-block to fit it, then if all is well, you can cut out the profile of the link and finish it off. If not, then junk the metal and start again. The simplest way for beginners to produce the slot is simply to mark it out carefully, drill a row of holes to remove most of the unwanted metal, then work carefully with a file up to the scribed lines.

File up the die-block and fit the two together, easing out the link until the block slides from end to end freely but without shake. A good craftsman exercise this, but if you want to machine the job, fine. Those with millers and rotary tables will feel smug here. Humble vertical-sliders can make up a suitable tool using a screw and nut arrangement to move the link material around a pivot at the right centre distance. This is presented to a suitable slot-drill and the curved slot progressively produced.

Expansion links and die-blocks are generally hardened to resist wear, and there are two routes to achieving this. The first is to use ordinary mild steel and case harden it. This is done by heating the link to red-heat and dipping it into case hardening compound. Essentially this is just a form of carbon, a small amount of which 'dissolves' in the hot iron on the surface to produce a thin skin of high-carbon steel which, if quenched from red-heat in water, will become glass hard. Successive heatings and dippings increase the depth of the hard layer, but two or three goes are sufficient. The advantage of case hardening as a process is that the core metal is unaffected and hence the component does not become brittle. The disadvantage is that it can be a rotten job cleaning up again to bright metal. The preferred material for expansion link is probably so-called gauge plate. This is a high carbon steel produced in accurately ground slabs, in the soft state. A link cut from this material can be hardened by heating and quenching just once, making the process that much easier. Use the same stuff for the die-block and harden that out as well.

The components for the reverser stand present some awkward shapes for manufacture, but a systematic approach and some careful filing will produce the

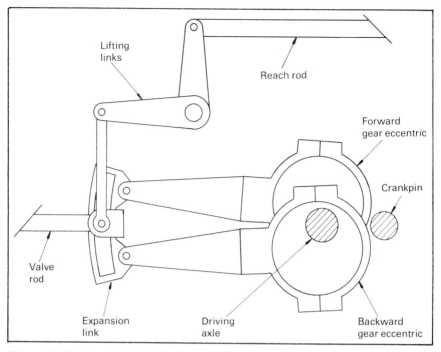

Figure 7.2: The layout and principal components of Stephenson's valve-gear.

right results. The notches in the sector plate are not cut until the gear has been assembled (see Figure 1.4).

Setting up and timing Stephenson's valve-gear calls for some patience and care but is essentially fairly simple, if the components have been reasonably well made. The problem is that one is generally poking about between the frames with not much space for the fingers, however the procedure is roughly as follows. First of all set the eccentrics in approximately the right position — your drawings will probably show this, but simple experimentation will suffice. Do not tighten the grub screws down hard at this stage. Next, put the cab reverser fully forward and turn the loco wheels gently in the forward direction with the fingers. This is probably easier said than done, since the whole thing is likely to be on the stiff side, and also rather oily; however, be persistent! Watch the die-blocks and see if they hit the end of the slot in the expansion links as the wheels revolve. If they do, then the cab reverser is too far forward, so move it back a little and try again. Carry on until the reverser is as far forward as possible without the die-blocks actually touching the end of the slots, then cut a notch in the sector plate so that the reverser can be latched into this position. Hopefully the two sides of the valve-gear, left and right, are symmetrical, but check both die-blocks to be sure that neither is clouting the expansion link. Now, repeat the process with the cab

reverser or 'pole' fully backwards whilst turning the wheels backwards. This then establishes the full gear positions for both forward and reverse, and, if you have been peering at a tiny die-block deep between the loco frames, it probably establishes you as an outside valve-gear fan.

Now we turn our attention to the slide valve and steam ports, removing the steam chest cover to see what is going on inside. Put the reverser in full forward gear, then adjust the position of the valve on the valve rod by whatever method your design allows, so that the steam ports open equally at each end of the valve travel. Next, slacken off the forward eccentric of the cylinder you are working on so that it can be pushed around the axle but will stay where it is put — this will involve removing the back of the eccentric strap. Put the crosshead on front dead centre and turn the forward eccentric in the forward direction, until the front steam port is just about to open. Tighten the eccentric grub screw fairly well and then repeat the operation on the backward eccentric, with the reverser in full back gear and whilst turning the wheels backwards. A simple enough operation but rather fiddling at times. Now do the same for the other cylinder and you will have a reasonably tuned set of Stephenson's valve-gear.

It is probably a good plan at this stage to make up the steam pipes (described later in this chapter) and rig up a compressed air supply so that you can run the chassis and check that all is well (described later in this chapter). After the run, clean up and tighten down all the fixings properly. Cut another notch in the reverser sector plate half way between the other two to give the mid-gear position, then cut two or three notches between mid- and full-gear in both directions to give some intermediate gear positions.

Walschaert's valve-gear

Walschaert's valve-gear was probably the most successful of gears in full size; certainly it was the most common, and so it appears frequently on models. It is similar to Stephenson's gear in having a curved expansion link for reversing, but there are some subtle differences. Broadly speaking, Walschaert's gear has two separate systems which combine to drive the valve. The first part is fairly easy to understand. The fundamental component of this system is the expansion link which, unlike the link in Stephenson's gear, is mounted on trunnions so that it can oscillate to and fro. It is made to operate thus by the return crank on the driving crankpin, via the eccentric rod. In the case of inside Walschaert's gear, the motion is derived from an eccentric on the driving axle. In common with Stephenson's gear the expansion link carries a die-block which, in partnership with the other part of the gear system, drives the valve. This die-block can be slid up and down the expansion link using the cab reverser, so that the engine will run in one direction when the die-block is above the trunnions and the other direction when it is below. The second part of the system is more cunning in its operation but mechanically it is deceptively simple. If the valve had no lap, the expansion link system would do all that was necessary, but because the valve must travel further in order to crack the steam port on dead centre, some extra motion must be introduced. This motion is derived from the crosshead via the drop arm and union link and is mixed with the motion from the expansion link by the aptly-named combination lever. Consider now what happens when the loco is in mid-gear, that is, with the die-block in line with the expansion link trunnions. As the wheels are

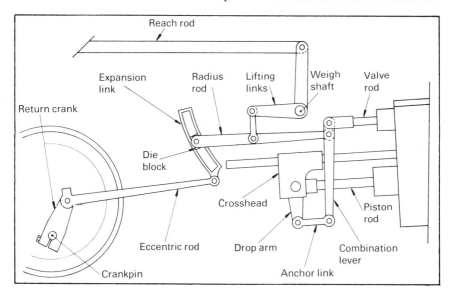

turned, no motion is transmitted to the combination lever by the return crank, so the joint of the combination lever and radius rod stays put. The crosshead, however, moves to and fro, taking the bottom end of the combination lever with it, so that the combination lever pivots about the end of the radius rod, and hence moves the valve to and fro by a small amount. This small motion is sufficient to crack open the steam ports at each dead centre. Incidentally, we are considering only outside admission slide valves, as found on models, here. With inside admission piston valves, the radius rod joins the combination lever above the valve rod, but I suggest you refer to some of the more advanced books to investigate this aspect.

Now if you set the crosshead on one or other of the dead centres, you will find that the die-block can be moved from end to end of the expansion link without moving the valve rod, because the expansion link is in exactly mid-travel. What all this means is that the steam port will crack open on dead centre, whatever the position of the die-block in the expansion link.

The amount by which the port cracks on dead centre is known as the lead, and this will be constant whatever the position of the die-block: forward, reverse or anywhere in between. The idea of giving the valve timing some lead is to give the steam a chance to build up pressure on the piston head early on in the stroke and not later on when the valve is about to close up again anyway. This gives the piston a good hard thrust of high pressure steam which can then work expansively for the duration of the piston stroke, using the steam economically. That, at any rate, is the theory; whether it works out in model practice is debatable, since the lost motion in any but the finest valve-gear is likely to be similar in magnitude to the theoretical optimum lead. Even so, we may as well stick to theory, since it is easier to check valve events visually when the port is cracked at dead centres.

Having cracked the port at the correct time, on dead centre, the distance that the valve then travels is governed by the position of the die-block in the expansion link, in turn controlled by the cab reverser. In full gear the valve travels a long way and steam is cut off late in the stroke. As with Stephenson's gear this might be about 75 per cent of the stroke. As the cab reverser is moved towards mid-gear, the valve travel lessens, and steam is cut off earlier in the stroke to give 'expansive working' as it is called, with consequent economical and smooth running.

A further benefit from this clever system is that the valve changes direction abruptly just before the end of the stroke and so quickly opens the steam port to exhaust via the cavity in the slide valve. This results in a nice sharp puff of exhaust from the blast pipe to draw air through the fire in a brisk manner.

Thus it is that the expansion link system controls the major travel of the valve, whilst the crosshead system makes sure that the steam port is cracked on dead centre whatever the direction of travel of the loco, or the degree of cut-off. The two are put together at the combination lever to give a near-ideal sequence of valve events with a remarkably simple system of levers. Just for those with a mathematical turn of mind, the Walschaert's gear combines the sine and cosine of the phase angle, so now you know!

The problems of manufacturing the components for Walschaert's gear are

Walschaert's valve-gear on a nice 5 in-gauge 'K1' loco with lots of extra scale detail.

similar to those for Stephenson's, although Walschaert's gear has a couple of extra components to make. The main worry for beginners will be the expansion link with its trunnions and associated motion bracket, all of which are a bit fiddling, particularly in smaller gauges. Some designs of the Walschaert's expansion link are very complex, with box structures and twin die-blocks, but the simpler model designs are usually made up from a single flat plate with a suitable bracket, carrying the trunnions, brazed on.

One problem I always find with small brazed assemblies such as these is the cleaning up afterwards. Somehow it is hard to get everything back to nice, bright metal from the blackened state that results from heating. One possible solution here might be vacuum brazing. This is a specialist process, and you will need to go through the *Yellow Pages* and find a company that does it, but there are several jobs that might be tackled in this way so the effort could be justified. One component in particular comes to mind here, and that is the crank-axle of an inside-cylindered loco.

To describe vacuum brazing briefly; the components to be joined are coated with a paste of suitable brazing material and fitted together, then placed in a vacuum furnace. All air is then pumped out and the oven heated electrically to a suitable temperature, cooled and opened to atmosphere again. No flux is needed since there is no possibility of oxide forming, and the components very often come out cleaner than when they went in! Distortion is minimal since the items are heated steadily over a period of time, held at a controlled temperature and let down gradually. The pieces do have to be held in place during processing, but then the same is true of conventional brazing generally.

As for the rest of the Walschaert's gear components, it is a question of milling, sawing, drilling, filing and finishing until you have the required bits and pieces. It is worth prolonging the life of the valve-gear by bronze bushing where possible or case-hardening otherwise. In the latter event the old problem of blackening crops up again, and it might be worth having a word with a specialist heat treatment company, since they have all sorts of cunning modern processes such as induction hardening. You may well find that they also do vacuum brazing.

Setting up and timing Walschaert's gear is pretty easy since most of the events are defined by the geometry, however, the return crank needs to be set accurately and the length of the eccentric rod is usually established at the same time. The procedure is as follows; set the crosshead on front dead centre and the expansion link in mid-gear, that is so that the die-block can be run up and down without moving the valve rod. Put the return crank roughly in place, according to your drawings; this will probably be about 90° ahead of the main crankpin. Now take a pair of dividers and set them to the distance between the hole in the expansion link tail and the return crank small end. Then move the crosshead to back dead centre and check the measurement again. Move the return crank around the main crankpin until the distance between the expansion link tail and the return crank small end is the same for front and back dead centres. Clamp the return crank at this setting and, ideally, fit a taper pin to prevent it shifting in service. In desperation a friend of mine welded the return cranks of his loco, but regretted this when it came to overhauling at a later date! Make up the eccentric rod to the length eventually derived using the dividers, and repeat the whole job on the other side of the engine.

You next little job is to set the valves, and this is pretty simple provided that the design allows you to adjust the position of the valve on the valve rod easily. It is simply a question of removing the steam chest cover so that the ports and valve are visible, then rotating the wheels gently by hand. The valve should just crack the steam ports by the same amount at each dead centre. If it does not, then move the valve along the rod until it does, then fix in place according to your design, and reassemble the steam chest for good and all. All being well, it will be your grandchildren who next dismantle these parts for overhaul, somewhere around the middle of the next century!

The final job is to cut the notches in the reverser sector plate, as with Stephenson's gear. In this case put the lever fully forward and rotate the wheels forwards by hand. This will probably cause the lever to kick back slightly, so move it back little by little until no kick is evident as the wheels are turned, then cut a notch for the latch at this position. Do the same for reverse, with the lever fully backwards whilst turning the wheels backwards. Cut a notch in between for mid-gear and then make a suitable number of intermediate notches to provide for a variety of cut-off positions. If your engine has a screw reverser, then substitute marks on the indicator plate for notches.

If all is well, the loco chassis should now turn reasonably easily in either direction, and the valve-gear should run from full forward to full reverse without tight spots. Naturally the packed glands and new joints will be fairly stiff to begin with, indeed, it may not be easy to turn the wheels with the fingers at all, however, provided that the resistance is fairly even, this is not too serious. Things will tend to free up later on. If, however, the chassis has marked stiff points and free points as the wheels are turned, then some detective work is probably called for. This will involve identifying and easing the guilty components to allow freer running, but go carefully and only remove the absolute minimum of metal. This can be rather depressing for the beginner who finds that, for all his efforts, he is rewarded with a mechanism that runs with all the smoothness of a three-legged camel, however, do not be discouraged. I found that my first loco freed up very quickly despite being 'notchy', and continues to run nicely after some fairly heavy work. In fact, steam engines are crude devices, particularly in comparison with modern petrol engines, and will work quite well with gross errors and shocking fits, as anyone who witnessed the railways in this country towards the end of the last war will know.

Steam and exhaust pipes

The pipework connections to the cylinders should perhaps come under the heading of either 'boiler fittings' or possibly 'plumbing', but they are usually considered as part of the chassis group. One reason for this is accessibility, since it is hard to assemble these bits if the boiler and smokebox are already fitted. Another good reason is that it allows the chassis to be air tested, an exciting stage in the construction of a model loco! The steam and exhaust pipes are relatively simple components, but there are some points to watch for. The standard arrangement, particularly in the smaller gauges, is to use a central T-piece, often in the form of a casting. This has screwed connections to the cylinders and a union on the top in the case of the steam pipe, or a specially shaped blast cap in the case of the exhaust. This is fine, but in order to assemble this design, the pipes going to the cylinders have to be screwed deep into the T-piece in order to lower the unit between the frames. After this they are then screwed out again, into the cylinders, and secured by locknuts. This arrangement is hardly ideal for sealing against full steam pressure, and I suspect that most of the time it is the locknuts alone which provide the seal. Clearly, fairly close fitting threads are desirable from the steam-tightness point of view, but, on the other hand, this then demands that four sections of thread, two in the cylinders and two in the T-piece, are all in line.

Plumbing work for a 7¼ in-gauge loco. On the right the exhaust pipe complete with blast cap incorporating blower union. On the left is the steam pipe with unions for lubricating oil supply.

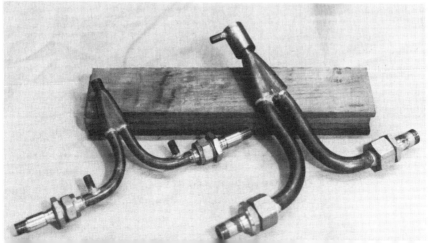

Something of a dilemma, however a good helping of plumbers' jointing compound — a sort of runny putty — helps a lot and generally the system seems to work.

Larger scale designs may produce better engineered solutions such as flanged and bolted connections, but these may call for some rather tight bends in the pipework. Annealing the copper by heating to dull red and quenching is helpful, but the best solution is to borrow or make a small pipe-bender. Otherwise try using elbow connections made from castings or solid. I know that in theory clean, streamlined steamways are desirable, but a pound or two of backpressure sacrificed to make for easy, practical manufacture, is neither here nor there. Most designs will call for one or two connections in the steam pipe for oil from the lubricator and these will probably include a check valve. This is a simple ball-device, but take care that it seals well, because back flow of steam into the lubricator causes all sorts of puzzles with vapour locks and so forth.

Do not jam the blast cap down too tightly by the way, as you may want to change the nozzle diameter following steam trials. A final point here, it is important that the blast nozzle is concentric with the funnel, and points directly up it, so take care that the blast pipe is central and vertical, and that in subsequent work on the chassis it stays that way.

The lubricator and air testing

Adequate lubrication of the slide valves and cylinders of a steam loco is essential in full size and model practice, otherwise seals will be non-existent and wear will be catastrophically rapid. The answer is to pump oil into the main steam pipe ahead of the steam chests, and let the flow of steam blow the oil in atomized form through to the slide valves and on into the cylinders. It then, of course, proceeds up the blast pipe along with the exhaust steam and emerges at the funnel. A rim of black oil around the funnel of a working model indicates that the lubrication system is working as it should. Unfortunately, this oil tends to spatter the engine, driver and train, but at least it is not the castor oil of World War 1 aircraft engines with the consequent side effects! Indeed, the oil used is specially formulated for use with super-heated steam, and is black and treacly when cold.

Cylinder lubrication presents two problems; one, to get the oil into the steam pipe against full steam pressure, and two, to introduce it at the right sort of rate. Early model locos, and indeed, current stationary engines, carried displacement lubricators which are essentially oil reservoirs in the steam pipe. A certain amount of steam condensed in this reservoir on its way through to the cylinder, and the resulting condensate sank to the bottom of the tank, displacing a corresponding volume of oil into the steam way. This works after a fashion but is rather hit and miss, and tends to gulp oil as the regulator is closed. It is used in a more sophisticated form known as 'sight feed lubrication', but the majority of model designs now have a small pump housed in a tank of oil, and driven by a ratchet mechanism off some part of the valve-gear or water pump.

The oil pump itself will probably be one of two possible designs. LBSC favoured an oscillating cylinder pump, whilst later designs have a fixed cylinder

Above *Typical oscillating pump lubricator, this one is for a 'Rob-Roy' tank engine.*
Below *This 'Rob-Roy' has been test run successfully on compressed air and is all ready for its boiler.*

and a piston with a large flat disc at the top end, depressed by an eccentric and returned by a spring. In both cases the drive is by a lever and ratchet mechanism arranged so that each revolution of the wheels turns the pump crankshaft by only one or two teeth of the ratchet wheel. Thus there might be thirty or forty wheel revolutions to each complete stroke of the pump so that the correct quantity of oil can be steadily fed to the main steam pipe. The size of the pump ram is around about $\frac{1}{8}$ in in most $3\frac{1}{2}$ in-gauge designs, so that the whole affair is pretty small, the oil tank containing the pump being typically 1 in square in plan and $1\frac{1}{2}$ in high. The lever and ratchet mechanism may consist of a suitable ratchet wheel equipped with two pawls, one on the lever, the other on the body, or it may use small ball-locking clutches which are available from some model suppliers. The pump assembly will carry a non-return ball valve at the delivery, and as mentioned another check valve is usually fitted up close on the steam pipe. This set-up prevents steam leaking back into the oil tank and causing problems. All of this may sound rather small and tricky to make, and you may well chicken out and buy a ready-made lubricator for your engine, however, they are really not so difficult to make, particularly if you buy at least the ratchet wheel ready-made.

Lubricators are often sited between the frames, just behind the front buffer beam. This is nice and close to the steam pipe, and is hidden away out of sight thus improving the scale looks of the model, but apart from these points it is a rotten place for the thing. Not only is it inaccessible to fill and generally to maintain and repair, but it will inevitably collect up grit from the smokebox when the flue tubes are brushed. Generally, you can fit the lubricator anywhere that a suitable oscillating drive is available, but a good site, as used on 'Rob-Roy', is up on the running boards. The lubricator is then easily and cleanly accessible, and can be driven by the crosshead or perhaps the expansion link. It may seem a little out of place here, but it can be disguised to look like the toolboxes or sandboxes often carried on engines in this particular place. Indeed, some full size locos carried Wakefield lubricators in just this manner.

At this stage, with valve-gear, steam pipes and lubricator assembled, the chassis is ready for an exciting time, that is the first run on compressed air. It is well worth obtaining the use of a compressor and rigging a connection to steam pipe for two reasons. Firstly, it is a chance to check that all is well, and make adjustments if necessary, before access to the works is impeded by the boiler and platework. Secondly, and perhaps more importantly in many ways, there is the psychological boost a running chassis will give to the project as a whole.

An air supply of as little as 20 psi will probably be adequate, but a standard 100 psi air line is best, provided that some control is available to adjust the delivery pressure. Fix up some form of union to the steam pipe and then raise the chassis on blocks or make a carrier as shown in the photograph of *Lady Caroline* in Chapter 3. Oil all the other working parts thoroughly with machine oil, put into full forward gear, then open up the air supply a little way. If the fates are on your side, the chassis may spin into life immediately, but it probably will not. Ease the wheel treads round carefully with your fingers, increase air pressure and you will no doubt be in business. Do watch out for your fingers because the chassis may start suddenly and with quite a lot of power behind it, enough indeed to do quite a lot of damage. Do not let the chassis run too fast, because it is running light and may be able to go fast enough to do itself some injury. It will probably be rather notchy to begin with, but do not worry too much, this will likely even out during the 'running in' period. Stop the chassis and run it in reverse for a time, then try notching up towards mid-gear and see how things go. There should be a nice steady purr from the blast pipe, with all the exhaust beats even, but if you can detect a continuous blow at low speed, then one or both of the valves is not seating correctly on the port face and you must investigate. If the worst comes to the worst and it will not run at all, check all your connections and settings, then find someone from the local model club to help. An experienced loco man may sort you out pretty quickly, with luck.

Once the chassis is behaving itself well, clean up the more obvious oil splashes, tidy the workshop and call in your family, friends, neighbours and anyone passing, to witness your amazing skill and patience. After this, store the chassis somewhere where it will not be prone to rust, preferably in the house, and start thinking about the next stage, the boiler.

Chapter 8

The hot bit

I have given you some idea of boiler design in Chapter 1, but we should perhaps talk now about the various designs of boiler the model locomotive man will encounter. These are largely based on the boilers that evolved in full-size practice. Incidentally, we are going to deal only with true fire-tube locomotive boilers, not the various water-tube designs that have been applied to models in the past. Those are now quite obsolete in the passenger-hauling gauges and I think we can safely ignore them. The simplest type of locomotive boiler has a parallel barrel, a round-topped firebox, which on the outside is just an extension of the barrel, and the lower part of the firebox is narrow enough to fit down between the frames. See Figure 8.1.

This design is found on the smaller boilered version of LBSC's 'Tich'. As steam requirements become greater, then a larger diameter boiler barrel is needed and we get to the situation in Figure 8.2 in which the sides of the firebox are brought inwards from the full width of the barrel to accommodate the firebox between the frames. 'William' and 'Springbok' by Martin Evans have boilers like this.

Most heat transfer from the fire to the water in the boiler takes place around the top of the firebox and it seems logical therefore to make the most of this area and so the next stage in boiler development is the so-called 'Belpaire' firebox as shown in Figure 8.3. In common with the Walschaert's valve-gear, by the way, the Belpaire firebox was invented by a Belgian. Very often the barrel of a Belpaire boiler will also be tapered, narrowing towards the smokebox, making further complication for the boiler builder. Worse still, this taper is often asymmetric. The bottom edge is then parallel with the axis of the locomotive and the top edge slopes down towards the front. Whilst tapered barrels can be made from a piece of solid drawn tube, this is not easy so they are best rolled and seamed by the builder from sheet.

In order to accommodate a big fire, the final stage in boiler development was the wide firebox or 'Wooten' firebox as named after its inventor. See Figure 8.4. This spread beyond the width of the frames, indeed right out to the maximum width of the engine, resulting in a huge grate area for the hapless fireman to feed with coal. The frames then had to be engineered to pass under the firebox to the back of the engine, often carrying a pair of trailing wheels. Good examples of this were locomotives like Gresley's 'A3s' and 'A4s', and Stanier's 'Duchesses'. These had sloping backheads and throat plates and the inner firebox was extended forward

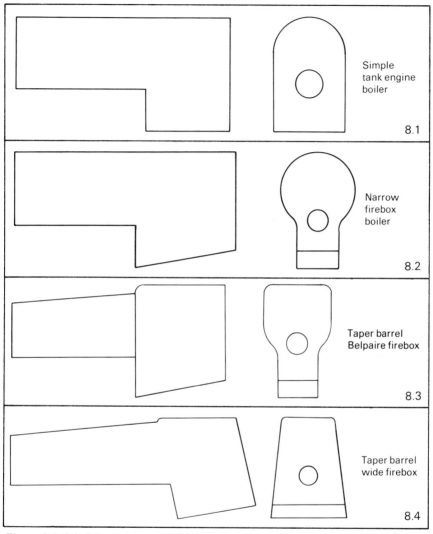

Figures 8.1, 8.2, 8.3 and 8.4: Some of the types of boiler found on steam locomotives. Side and rear views.

into a so-called 'combustion chamber' which reduced the length of the boiler flue tubes to some advantage.

All these features make for a pretty complex boiler requiring some skilled work, so the beginner would be well advised to stick with a design having a parallel-barrel, round-topped, narrow firebox, as in Figure 8.1 or 8.2.

You do not, of course, necessarily have to build your own boiler. Quite a

Above *This boiler was made by a professional and features a Belpaire firebox and tapered barrel.*

Below *The professional at work. This is Denis Bishop of Bishop Richardson Model Engineers, Birmingham. Here he is warming the boiler with propane prior to soldering with oxy-acetlyene. Note the very simple set up.*

number of firms advertise professional boiler building in the model press and once again the cheque book can by-pass a lot of hard work. Prices may appear high to the uninitiated but in fact they are not unreasonable. They range from about £100 for a 'Tich' or 'Rob-Roy' boiler to £800 or £900 for a 5 in-gauge 'Britannia' or 'Duchess' boiler, but something like three-quarters of this will be in raw materials alone. Copper is moderately expensive and silver solder, with its high silver content, must be very expensive by anybody's standards. Thus one actually buys the labour of construction quite cheaply and given that the home constructor will need to obtain various bits and pieces of equipment into the bargain (see next section) then direct purchase has much to recommend it. One small point may be worth mentioning here. Any reasonable boiler maker will test his own products and issue a certificate, however, not all model clubs will accept this for insurance purposes and may ask to witness a test themselves. In order to save yourself work, ask the supplier to let you have the bungs and caps he used to test the boiler, then you can turn the job over to the local 'boiler-busters' with the minimum of effort.

None the less, there is much pleasure to be gained from this aspect of locomotive construction and there is no reason why a complete novice cannot make a good job of making his own boiler. I propose to give an outline of boiler building in this chapter, and beginners must supplement this information by reading some of the more advanced books and articles in the model press.

By the same token, this is another good time to seek advice, if not actual help, from members of the local model engineering society. If someone with experience of boiler building can look over your shoulder, particularly during the early stages, then life will probably be a lot easier and the final job much better. Indeed it may be possible to borrow not only experience, but tools and equipment for the job. Advice on where to lay hands on materials is also useful for the novice.

I think it should also be said, at this stage, that boiler construction has something of a black art about it. Precision engineers perhaps feel that hammering out metal, heating, quenching and steaming vats of acid are not quite in keeping with their skills. Whatever the reason, people seem less willing to tackle a simple boiler, than to tackle a complex chassis. The result is the proliferation of people offering ready-made boilers. There is nothing secret about boiler making and although, like any other activity, some will be better at it than others, if you fancy trying your hand, go ahead. It does mean that when the admiring and incredulous public ask you if you built the locomotive yourself, you can honestly say 'yes' and not lead off into half-hearted excuses about not having the time or the equipment to do the boiler! You will also be able to enter competitions, both static and working; and do not think that first locomotives do not win prizes – they do.

Materials, processes and tools

There are several materials from which model locomotive boilers might be made, but when all is said and done, copper is by far the best material. Mild steel was used for full size boilers, quite often in conjunction with copper for the flue tubes and inner firebox, indeed steel can be used for models with some degree of success. It is a fraction of the cost of copper and very much stronger in terms of absolute tensile strength, but suffers from the very major disadvantage of corrosion. Steps can be taken to minimize this problem by using soft water and always completely filling the boiler with water at the end of a run, but by and large, corrosion scares off most builders. A further disadvantage of steel for model boilers is the relatively poor heat conduction of the metal which will inevitably reduce the efficiency of steam raising. It is not easy to flange steel boiler plates so that welding will be required, calling for a moderately skilled craftsman if the joints are to be safe. Despite all this, perfectly adequate steel boilers have been built for 5 in-gauge and there is indeed a lot to be said for steel. Some clubs may demand more frequent testing for steel boilers, say every year instead of two years, and this does not seem unreasonable.

Another possible material for boilers is stainless steel, which largely gets round the problem of corrosion, however, it has its own peculiar disadvantages. Firstly it is pretty well the same price as copper, and secondly it is rotten stuff to work. Hacking out boiler plates from $\frac{3}{16}$ in stainless steel is my idea of purgatory and welding, which is probably the best form of assembly, will involve some fairly exotic process such as TIG (tungsten-inert gas) welding. A further, and perhaps more obscure problem with stainless steel is known as chloride stress corrosion cracking or CSCC. This is a form of cracking which takes place in stainless steel under stress where chloride ions are present, particularly at high temperatures. It can cause the failure of a stainless boiler after a depressingly short period of service, and since it is quite hard to obtain quantities of chloride-free water, we had better leave stainless steel alone for boilers. Indeed, its use on a locomotive is best confined to components like piston and valve rods. Stainless buffer heads may look sweet but they are hardly scale practice. What is a chloride ion I hear someone ask — well, it is a chloride atom that has gained an extra electron, giving it a negative electrical charge. When common salt, sodium chloride, dissolves in water it disassociates into sodium ions and chloride ions, that is, the sodium atoms lose an electron and become positively charged (electrons carry a negative charge) whilst the chloride atoms snap up these wandering electrons to become negatively charged. These charged atoms are called ions. Here endeth the first lesson in solution chemistry.

This then leaves us with copper, which is an excellent material for model boilers. It has very high thermal conductivity, it resists corrosion, it can be formed into flanged plates and can then, quite easily, be soldered together. Cost is a disadvantage but with careful use, your copper boiler will probably outlive you, so it is a once for all investment.

In full size practice, riveting was the accepted form of boiler construction replaced later on in the steam era by welding, and indeed there is a parallel with model boilers. The earlier fire-tube model boilers were held together by close riveting all the seams, followed by caulking with soft solder. This worked, after a fashion, but it was pretty difficult to get rivets into some seamed areas and the caulking often leaked as a result. Strength-wise these boilers also left a lot to be desired and they were fairly soon succeeded by hard soldered types, with soft solder reserved only for the caulking of the stay-heads (see later).

The most basic of hard solders for copper is brass, however, brazing requires rather high temperatures and brass filler does not run freely into joints, so the standard material for the job is now silver solder. There are a number of grades or types of silver solder with a range of melting points, but, they all contain a fairly high percentage of silver, which gives the twin desirable features of low melting point and free running. Hence the trade names used by Johnson Matthey, the main producer of silver solders — 'Easyflo' and 'Silverflo'. The composition and melting range of these solders are given in the table.

Silver solders, composition and melting ranges

| Name | Nominal composition | Melting range in °C | |
		Solidus	Liquidus
Easyflo	50 per cent Ag Cu Cd Zn	620	630
Easyflo No 2	42 per cent Ag Cu Cd Zn	608	617
Silverflo 24	24 per cent Ag Cu Zn	740	780
Silverflo 16	16 per cent Ag Cu Zn	790	830

Data by courtesy of Johnson Matthey Metals Ltd.

A prudent scheme for assembling a boiler is to use a higher melting point solder for the initial joints, such as the main barrel seam and the bushes which carry the boiler fittings. The subsequent joints can then be made with a lower melting point material, hence reducing the risk of softening the previous ones. In practice this is not so much of a problem due to the effect of solvation. This is the result of the parent copper dissolving to some extent in the melted silver solder in a joint. The new alloy thus formed has a higher melting point than the original silver solder with the result that a boiler can quite easily be made using one solder throughout. The usual choice for this is 'Easyflo No 2'. Even so, the beginner might do well to put the bushes into the backhead and barrel with 'Silverflo 24' and then use 'Easyflo No 2' for the rest of the work.

A point worth noting at this juncture is the high percentage of cadmium in some of these solders, and the environmentally aware will know of the toxic properties of this metal. Industrial use of silver solder demands fume extraction at the point of use, but as far as we amateurs are concerned a gentle draught away from the operator is probably enough, but avoid getting close to the work and taking a lungful of the fumes. You will not drop dead on the spot, but cadmium has a cumulative effect on the central nervous system and frankly, we all need every brain cell we can muster these days! The alternative is to use one of the cadmium-free solders such as 'Silverflo 55' and this is the course of action recommended by the manufacturers, however, this material does not run as freely as 'Easyflo'.

The vital adjunct to silver solder, indeed all solders, is some form of flux, and this is one of those items bolstering the black-art image of boiler making. Its very name conjures up images of eighteenth century alchemy. No magic here though, just some more solution chemistry. The philosophy behind soldering is that the solder should actually alloy itself with the surface metal of the components to be joined so that, on an atomic scale, there is a graduation from parent metal to solder and back to parent metal, across the joint. The fly in the chemical ointment is atmospheric oxygen which produces a layer of oxide on the metal, especially when it is heated, preventing the alloying process from taking place. The answer here is the use of a flux which is a solvent for the oxide, thus removing oxide as quickly as it forms. No problems, except that once the flux is saturated with oxide it is useless, so in general one should keep heating times to a minimum and flux coverage to a maximum.

Quite a number of compounds will act as fluxes, borax being the classic one for brazing and even common salt is possible for use with iron but it has a very high melting point. I have, on my shelves, a book, dated 1955, which suggests the use of potassium cyanide as a flux, which seems singularly foolhardy. The best plan for us is to use the flux provided by the 'Easyflo' manufacturers, and imaginatively entitled 'Easyflo flux'. This is a cocktail of potassium, boron and fluorine salts designed to give optimum effect.

If you have not used silver solder before, have a go on some bits of brass to begin with. The rules are as follows:-

Cleanliness   The metal of the joint must be as clean as possible which means bright metal, freshly machined, filed or emery-clothed at the time of soldering. Ideally, the components should fit with a gap of 2 thou or so for the solder to fill.

Flux   Mix a suitable quantity of flux powder with water to the consistency of cream, and apply liberally to the faces to be joined, then fit them together and apply more flux from the outside. Use only freshly mixed flux paste.

Heat — Stage 1   Warm the job steadily with your torch (see later) so that the flux paste boils and drives off the water. Bring up the temperature all round so that the flux then fuses into a clear liquid all around the joint.

Heat — Stage 2   Now heat the job more strongly, concentrating on the areas of greater metal mass since these will heat more slowly.

Solder   As the job comes up to dull red heat, dip the strip of solder in the flux paste and touch the job. If it is hot enough, the strip will quickly melt and 'flash' round into the joint, showing a characteristic silvery-gold line. Apply a touch more solder if it has not run right round the joint. Coax the solder round with a piece of iron wire, if necessary. Never attempt to melt the solder with the torch flame, the job will do the melting if the temperature is right. Do not apply excess solder, it is runny stuff and will simply run out of the joint and be wasted. As long as it shows all around the joint, then enough has been applied.

Cooling   Allow the job to cool for a while in air, a minute or so for a small job, five or six for a large one, and then quench. Use water for ferrous joints and pickle (see later) for cuprous ones. This will crack off the glazed flux and make cleaning up easier.

Cleaning up   Scrub with a stiff brush and then use emery or whatever is appropriate. In the case of boiler work, a scrubbing is enough.

This all sounds rather complex but after a couple of goes everything will fall into place and you will find yourself repairing all sorts of bits and pieces around the house and garden with silver solder!

Few tools are needed for boiler soldering but the source of heat is of prime importance. The best equipment for the expert is probably oxy-acetylene and indeed most professionals use it, but it is not the right thing for the beginner. A few years ago, the paraffin blowlamp was popular, but most people will now use bottled propane (note propane, not butane) gas (such as Calor Gas) and a suitable torch. You can, if you wish, spend a fortune on a complete set of burner equipment, such as that made by the Sievert Company, but a good alternative for the beginner is to rent the equipment for a week or so to complete the job. Most tool hire companies carry propane burners for jobs like paint stripping and charges are quite modest. Ask for a burner of about 70 ounces per hour gas consumption ($1\frac{1}{4}$ in diameter, part no 2943 in the case of Sievert equipment) for the average boiler. Take a cylinder of at least 15 lb capacity and make sure it is reasonably full because it is terminally frustrating to run out of gas part-way through the final heat up!

A large pair of tongs is very useful for moving the boiler around whilst it is hot, but large pliers and a pair of gloves may suffice. You will need some form of brazing hearth in order to operate at a comfortable height, perhaps of building blocks or thin sheet steel, something about 24 in by 18 in by 6 in. This should have firebricks at the back and side, obtainable from your local builders' merchant, and while you are there, try to get hold of some rock wool ('Kaowool'). This is an excellent insulating material and will withstand the heat of the torch pretty well. The boiler can then be packed in the stuff with only the relevant portion exposed, thus reducing heat loss by radiation. The previous alternative was to pack with coke, which is okay but not as clean and simple. Rock wool can also be used where asbestos was formerly recommended, such as the protection of the tubes whilst soldering the foundation ring.

Last, but not least, you will need a pickling tank containing a ten per cent sulphuric acid solution or one of the proprietary alternatives, large enough to immerse your boiler completely. The purpose of the pickle is to remove oxide from each assembly as it is soldered, preparing it for the next operation, a vital process in boiler work. A plastic container is best for the pickle, and a small plastic barrel is ideal. Sulphuric acid can be obtained from the model suppliers, but try your local garage or car battery dealer first. If you obtain concentrated acid then it will need dilution as follows. Work out the amounts of water and acid required and fill the container with clean, cold water. Set this swirling with a stirrer and add the acid is a steady stream, finally rinsing the acid container carefully into the pickle bath.

For some inexplicable reason the model press continues to pursue the myth that if water is added to acid, rather than the other way around, there will be some kind of catastrophic explosion. Budding terrorists might like to know that this is not true. What does happen when sulphuric acid is diluted, is that quite a lot of heat is liberated. If you add water to acid you can sometimes get localized boiling and hence spitting of the acid, not lethal but unpleasant. What is more important than the correct order of addition is that cold water be used. Incidentally, the reverse occurs with flux paste which goes cold when mixed with water. Put a lid over the pickle when not in use to prevent the cat falling in, and take care not to tip the container over, it makes a horrible mess of a concrete floor. How do you know when the pickle is exhausted? Simple, it will stop working and black oxide will not be removed from copper, by which time the solution will be a pleasant azure blue of copper sulphate. Watch out for splashes of pickle on your clothes, if you wash them out promptly no damage will be done, but if they are left overnight then holes will develop.

So, how do we go about using our copper and silver solder to form the relatively complex shape we need for providing the steam to run our engine? There are two stages in the process, the first is to make up the components from sheet material or pre-formed tube or bar, and the second is to solder them together tight against the forces of steam and water at a hundred or so pounds per square inch. Once again the prudent beginner will have chosen a design with the minimum complication, probably with a parallel barrel and round-topped narrow firebox, so I will go through a broad description of how this type of boiler is made, and leave the more difficult boilers, with their combustion chambers and thermic syphons, to the more advanced books.

The boiler components

The boiler tubes and flues will obviously be made from tubing as hopefully will be the main barrel. The type used should be 'solid drawn', that is formed without a seam, and of the thickness specified in your drawings. Actually, the fire tubes can be made thicker than is often specified in order to prolong the life of the boiler. These will then resist erosion by fast-moving ash particles with little effect on heat transfer. Large diameter tube for the barrel is expensive and a cheaper alternative is to roll it up from sheet using a set of bending rolls if you can lay hands on some, but then you will have to seam the barrel. This involves either a so-called coppersmith's joint, which is a lot of work for not much gain, or a butt strip brazed along the inside or outside of the barrel. An outside butt strip is preferable because then both you and the boiler inspector can see it, and it does not interfere with the flanges of the smokebox tubeplate and throatplate. Rolled barrels are fine, but the beginner had best stick to a drawn tube barrel if possible. Incidentally, the newcomer may hear talk amongst boiler makers of the need for oxygen-free copper, but this only applies where oxy-acetylene is to be used. This is because the very high temperatures involved can lead to brittleness in normal copper. Let the professionals worry about this one!

Perhaps the best job to tackle first, by way of components, is the flanged boiler plates, and the best way for the beginner to tackle these is with a cheque book. True, plate flanging is not particularly difficult, but it does involve quite a lot of work, and the price of a set of ready-flanged plates seems to me a bargain. You will probably find that the positions of all the tube and bush holes are marked out ready for drilling as well. I should also advise the beginner to buy a complete kit of materials along with the ready-flanged plates, as advertised by some of the model engineering suppliers. This will provide the odd bits and pieces such as bar for the foundation ring and thick wall tube for the firehole, which can otherwise be a nuisance to obtain in small quantities.

The reason for flanging the various plates is to provide a large soldered area to resist the sheer forces imposed by pressurization. If you must flange your own plates, then the procedure is to beat the flange out over a suitable former. These can be made from steel plate, hardwood, hardwood faced with thin steel plate or even cast in lead. A suitable copper blank is then thoroughly annealed and clamped to the former, after which it can be beaten gradually over the edge to produce the required flange. Anneal the plate again by heating to bright red and quenching in water the moment it shows signs of work hardening. You will need to anneal a number of times before the flange is fully formed. A hard plastic hammer is ideal for the beating operation and most people are surprised how easily the ductile soft copper takes up the required shape. However, even in the sort of simple boiler beginners will tackle, the required plates will be the smokebox tubeplate, throatplate, firebox tubeplate, firebox backplate and backhead. Perhaps you see now that ready-flanged plates are a godsend to the harassed novice!

Having formed the plates, there are a couple more operations to perform. Firstly, the smokebox tubeplate has to be turned in the lathe to fit the barrel, and the various tube and bush holes must be drilled. Drilling might indeed be adequate for the bush holes, but for the tubes we have a choice of approaches. Correct practice indicates that the tubes should have a clearance of 2 thou or so in the smokebox and firebox tubeplates for the solder to fill, however, this makes assembly difficult because they tend to fall through.

There are several possible solutions here, one is to support the ends of the tubes on a steel plate during soldering. Another idea is to produce an expanded ring about $\frac{1}{4}$ in from the end of each tube using a special compression tool. This process was described very nicely in the *Model Engineer* Volume 152, Number 3730 published in June 1984. The choice is yours but, when drilling and reaming tubeplates, work from the centre of the nest outwards to provide maximum support, or the holes may tear up. Lightly countersink the holes both sides so that the solder forms nice fillets.

The next job is to prepare the tube for the boiler barrel by squaring up the ends to length and for this a carrier will be needed. This consists of a suitable piece of studding, say $\frac{1}{2}$ in diameter, centred at one end for the tailstock centre, and carrying two wooden discs, held in place by nuts. These discs should be turned a tight push fit in the barrel tube and set an inch or so less than the finished barrel length apart. One end of the tube can then be held lightly in the chuck, the studding set on the tailstock centre, and the end of the tube turned true with careful light cuts. Reverse the tube on the carrier to deal with the other end. Whilst the tube is held true in the lathe, take the opportunity to mark the top centre line by fitting a scriber to the toolpost and traversing it down the tube. This will be handy for placing the dome, safety valves and steam turret later on.

In many boiler designs the barrel tube is cut and bent out to form the firebox outer wrapper, but generally the opened out tube does not provide for a deep enough firebox. The firebox sides must therefore be extended, but this will only be necessary on one side since one can bias the longitudinal cut to give one firebox side of full depth. The other side can be extended with an extra piece of sheet joined on with the help of a butt strip on the inside.

The firebox inner wrapper is formed from sheet, annealed and bent to shape. This can be a bit tricky, but careful use of fingers and a plastic mallet will produce a close fit on the firebox tubeplate and backplate. Now is the time to make up the crown stays which tie the roof of the firebox to the barrel and prevent its collapse under load. There are a number of types used, but most are made from sheet material, soldered to the firebox inner wrapper. Turn up the boiler tubes and flues to length in the lathe and then make up the various bushes which will be soldered into the boiler shell to accept the boiler fittings. The correct material for these is either drawn bronze or gunmetal, as supplied by the model engineering specialists, but not brass.

Many accounts of boiler building suggest that you should put in some rivets to hold the assemblies together prior to soldering. This is fine but things always have a habit of taking longer than expected, with the result that it may be quite some time from fitting a couple of components together, to getting around to soldering

Figure 8.5: This shows how a piece of seamless tube can be split to form the firebox outer wrapper, so that only one side needs an extension piece soldered on.

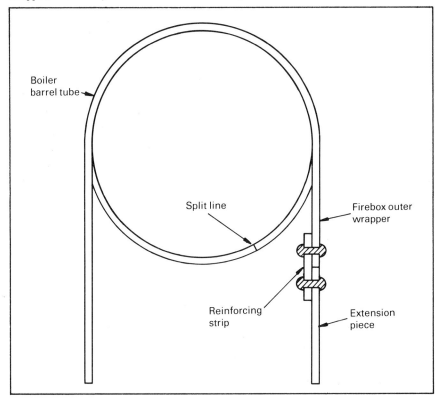

them. During this time oxide will form in the joint, dirt, grease and whatever else may get in and it will then be darn difficult to get the solder to take properly. The answer is to bolt the various assemblies together with a few small screws, and then, when you are all set to solder, whip the joint apart, rub with emery, apply flux paste and only then rivet up with copper rivets. No need for a posh job, just enough to hold up for soldering, after which the rivets can be filed off flush anyway. Having made up the main components of the boiler and prepared the tools and equipment, we are at last ready to start work on soldering it all together.

The first soldering operations

According to the saying, there are many ways to skin a cat, likewise with boilers there are a number of possible sequences for construction. You may have your own ideas, or wish to follow the instruction manual that accompanies your chosen design, but here is one possible sequence. Do not forget that we are talking about a simple boiler with a parallel barrel made from drawn tube and a round top narrow firebox. The first job then, is to solder the various bushes into the backhead, barrel and smokebox tubeplate. You can use 'Easyflo No 2' for these, or perhaps play safe and use 'Silverflo No 24', which has a slightly higher melting point. Incidentally, 'Silverflo 24' was, until recently, called 'C4', which is the name you will find in much of the literature. By the same token, 'Silverflo 16' used to be known as 'B6'.

The bushes should have 2 thou or so of clearance in their holes, which is a pretty rattly fit and something of a problem because we want the bushes true in the boiler, otherwise all the fittings will be cockeyed. This is particularly noticeable on things like the safety valves which look dreadful if they emulate the Leaning Tower of Pisa. Make sure the bushes sit nicely in their holes and then set the relevant section of boiler as level as possible before soldering. Make a start with the backhead, clean up all the joint areas to bright metal, apply flux paste and lie the plate horizontal on some rock wool or coke. Heat gently with your torch all over until the flux has fused to a clear liquid, then heat strongly until the metal begins to glow dull red. Dip your strip of solder in flux and touch one of the bush joints gently with the torch flame just clear, but still on the rest of the backhead. It should melt and run around the whole joint, if not, withdraw the solder and heat up once more, then try again until it does melt. Apply enough solder to run all round the bush and a touch more for luck, then move to the next. With an average size backhead, the subsequent bushes will probably be hot enough to take solder immediately so you can deal with them all in a few seconds. Once you are satisfied that all the bushes have had the correct treatment, turn off your torch. Wait a minute or so and drop the backhead into the pickle bath. Take care to avoid splashes of hot acid, in particular mind your eyes. Spectacle wearers will be at a natural advantage here, which is some compensation for those hefty opticians' charges! Leave the plate for five minutes or so, then fish it out and run it under the cold tap, giving it a scrub with an old nail brush if necessary. The backhead will now have a healthy salmon pink colour all over, and the solder will appear as a bright silver/gold colour. Examine each bush joint carefully to see that the solder,

or 'spelter' as the old coppersmiths call it, has run around each joint completely. Look particularly at the inside for a nice even fillet around the protruding bush. If there is any evidence of lack of solder or exessive pin-holes, then flux each bush again, both sides, and heat up again. You will probably have to heat pretty strongly to get the joints soft again, remember solvation, then apply more solder and scratch round with your scratch rod to aid penetration. That should take care of any problems, and you can go on to the smokebox tubeplate.

If we are going to heat up the barrel to take the feed clack valve bushes, and so forth, we may as well do the throatplate and firebox extension plate at the same time. Clean up all the parts ready, apply flux and rivet up the throatplate and firebox extension joint with a few rivets to hold things in place. Stand the boiler, barrel upwards, on the place where the backhead will eventually go, and pile rock wool up to the rear throatplate level. Apply some extra flux paste to the throat-plate joints and start heating up. The barrel will act as a heat sink and you will have to go at it fairly strongly to get the shell up to temperature, particularly if you have opted for 'Silverflo 24'. Once the whole thing is pretty hot, concentrate more on the bottom edge of one of the throatplate to firebox outer wrapper joints, with occasional wafts over the rest of the job. Start the solder running here and work along to meet the barrel, then work steadily round the barrel, forming a fillet in the angle between the barrel and the throatplate. Finally, work down the other throat-plate flange, making sure that there is plenty of solder at the corners where the flanges meet the barrel. This is where careful fitting of the components pays handsome dividends. If you find the gaps here just too large to fill with solder, do not panic, carry on and after pickling, make up suitable slivers of copper to plug them, and solder again. Having done the throatplate, allow the lot to cool a little so that the solder is not actually liquid still, then heave the boiler over on its side, with the firebox extension joint uppermost. Push rock wool around the boiler as best possible to conserve heat, and dollop some more flux on the joint. It will spit like a frightened cat but do not worry; then heat up again. Run solder well into the joint and do not forget to give each rivet head a touch also to seal it properly. If your design has blowdown valves (see later) at the base of the firebox, you can do one of these now. Next, move up to the front of the boiler, apply more flux to the feed clack bush and solder as before. Now turn the boiler on to its other side and deal with the opposite feed clack bush then stand it level on the firebox base and repack to reach the dome bush, safety valve bushes and steam turret bush. Cool, pickle, scrub and examine the shell carefully for less than perfect joints. If you find any, then flux and heat again. If any of the joints show signs of bubbling whilst soldering, use the scratch rod to release the entrained gas which is usually just water vapour.

One final job you can do with the 'Silverflo 24' at this stage is to solder the firehole ring into the firebox backplate, and then we can go on to 'Easyflo No 2' for the rest of the job. Do take care not to mix up the two solders in your store, or you will give yourself needless problems.

Top left *These are the two major sub-assemblies of a locomotive boiler. On the left are the firebox and firetubes, on the right the outer firebox and boiler barrel. In this case the boiler is a fairly complicated one with a Belpaire firebox.*

Right *The boiler shell with the front section of the foundation ring fitted, ready to accept the inner firebox assembly. The many holes are for the firebox side stays.*

Second soldering operations

The next phase of assembly, assuming you have at least paused for a cup of tea, is to put together the inner firebox and the tubes. Check that the firebox inner wrapper fits nicely on to the firebox tubeplate and backplate, clean up, apply flux, put in a few rivets and do the same with the firebox crown stays. Paste on extra flux and solder the lot together. This should be pretty easy because this is a relatively small assembly, but do not blast away at the tubeplate too much, or the thin webs between the tubeholes may start to burn away and thus ruin the job. Pickle and clean up and we are ready to put in the tubes. Stand the firebox on its backplate in the brazing hearth and flux the tubes also and poke them into the holes carefully, so that they all protrude $\frac{1}{4}$ in into the firebox. Make up some little rings of Easyflo wire and drop one around each tube to embed in the flux on top of the tubeplate. Fit the smokebox tubeplate at the other end of the nest of tubes to hold them all parallel, then adjust the whole set up so that the tubes are true to the firebox. Many designs have their tubes slightly inclined upwards towards the smokebox so remember to allow for this. Heat up the assembly steadily, concentrating more on the tubeplate because this is much thicker than the tubes and will take more heat to get it to temperature. Scratch each joint to make sure the

Left *The two assemblies have now been fitted together. In this example rod-type crown stays are to be used, and these have been pushed into place ready for silver soldering.*

Top right *The professional method; first the job is warmed all over with propane to glaze the flux.*

solder has run all round and add some more if necessary. Ideally, you should have a bright ring of solder round each tube, but a bright pool around the whole lot is no bad thing. Cool and pickle, knock off the smokebox tubeplate gently and examine the tube joints carefully. You should see a ring of solder on both sides of the tubeplate. If there are any doubts, cook it all up again, and then you will be ready to put the two sub-assemblies together.

Yet more hot work

At this stage things start to come together in both the metaphorical and literal senses, with the assembly of the firebox and tubes into the boiler shell. Those who have chosen to build a large engine will by now be thinking that something smaller or lighter would have been better. Lifting and moving 30 lb to 40 lb of near red-hot copper can be fairly strenuous exercise!

Start by sliding the firebox and tubes into the barrel, then fit the smokebox tubeplate part-way into the barrel and check that the firebox fits properly into the shell. It probably will not, so set to with some careful bending and easing of the crown stays and so forth until all is reasonably well. You may find it helpful to make up the front section of the foundation ring, the piece that goes in between the throatplate and the firebox tubeplate, and put in a couple of screws to hold the lot together, with the firebox central in the shell.

When you are satisfied with the fit, dismantle and withdraw the firebox assembly and get ready for the next big heat up. With more metal to bring up to temperature you will find that quite a bit of heat needs to go in at this stage. Clean all the parts to be soldered, that is, the smokebox tubeplate, front ends of the tubes, crown stays and the inside of the firebox outer wrapper where the crown stays will be soldered. If your design has girder stays on top of the firebox, then this last point does not apply. Flux the relevant surfaces and reassemble the boiler, pushing in the smokebox tubeplate, as shown in your drawings. This time, clamp the crown stays to the firebox outer wrapper with a couple of toolmaker's clamps, then drill through and rivet the stays to the wrapper. You will need a heavy bar of iron held in the vice and inserted into the boiler to act as a dolly to do this. Heave the boiler up on to the brazing hearth once more and stand on its back end, barrel upwards, then pile up rock wool or coke to near the level of the smokebox tubeplate. Plaster on yet more flux and put rings of Easyflo around the tubes, as before. Put a large one around the tubeplate and heat up. Have a good scratch round, add some more solder, as necessary, and, when the treatment is complete, shunt the boiler round on to its back with the firebox towards you to get at the crown stays. Pile up the insulation again, put on another dose of flux and lie pieces of Easyflo along each side of the crown stay flanges. Heat up, as usual, add more solder if the job looks at all dry, cool, pickle and scrub. Check that all the joints are sound and prepare for the final big heat-up.

The last major soldering operation is to fit the backhead and then the foundation ring. As a first step, make sure the backhead fits snugly into the boiler shell and over the firehole ring. It is probably a good idea to leave cutting the hole in the backhead for the firehole ring until now, when its position can be marked from the job and then cut and fitted accordingly. Many instructions for boiler builders suggest that the firehole ring should be peened (hammered like a rivet) over the backhead before soldering, but this strikes me as unnecessary, if not downright weakening. Leave a 2 thou gap around the ring and let the solder do its proper job. If the wrapper does not fit snugly round the backhead, drill some suitable holes through the wrapper and backhead flange, remove the backhead and tap, say 6BA, and clearance drill the wrapper. Use some bronze or home-made copper screws to pull the two together at the stage of fluxing and assembly. It does not matter what heads you use because you can file them off flush after soldering. With the backhead fitted but not soldered, make up the foundation ring, again looking for a snug fit all round. Use copper rivets or bronze screws to hold the pieces in place, pro tem. If you end up with any odd gaps, fill them with slivers of copper or copper wire, because Easyflo will not fill gaps wider than a few thou. Flux the lot before assembly, then set the boiler up in the hearth for the last blast. Best plan here is to cut a hole in the floor of the hearth to take the boiler barrel so that it can be rested on the throatplate, backhead uppermost. Pile up the wool or coke almost to the level of the backhead, and add extra flux. Heat up as usual and then work round the backhead joint, scratching and adding extra solder as needed to form a good sound job. Do not forget the firehole ring and give each screw or rivet head a dose of solder to seal it for keeps.

Allow the solder a moment to set then winch out the whole caboodle and set it on its back. Pile up the coke or wool and fill the firebox with it nearly to foundation ring level; this is to protect the tubes from the torch which is going to put a good bit of heat in to get the mass of copper around the foundation ring to dull red heat. Take care not to let coke dust get into the joints, however. Add more flux to the job and once again, work round the foundation ring with your scratcher and solder strip until all is sealed. Cool, pickle, scrub and check over for poor joints, then head for the pub for a celebration drink, because the heavy work is now over.

Stays and testing

The final operation on the boiler is to fit the stays, that is the ties which brace the flat surfaces of the boiler and stop them from bulging out under working pressure. Before doing this, however, it is well worth making a preliminary test for leaks using air pressure. Both the air test and the subsequent hydraulic test will need the boiler properly sealed, so now is the time to make up a set of plugs. Brass caps screwed in with PTFE tape (as used by plumbers) will sort out all the threaded bushes whilst holes for the dome and regulator will need cover plates, sealed by suitable gaskets. Leave one bush open and make up a special air-testing fitting to go into this. This fitting consists of a T in brass or copper tube, soft soldered together. The base of the T carries a nipple to fit the spare bush, on one arm of the T goes a pressure gauge, such as the one you will eventually use on the engine, and the other arm has a car tyre valve soldered in. Use a car tyre pump to put in 10 lb – 15 lb of air pressure, no more, and submerge the boiler in water. A tell-tale stream of bubbles means trouble, but do not despair. Mark the pin-hole, then dry the boiler, drill the hole and tap it, say 10BA. Then screw in a short length of copper wire, threaded 10BA, and chip off short. We can sweat over little plugs like

Oxy-acetylene is then used to solder each stay-head. The silver solder is applied here in the form of wire.

this when we are sealing up the stay-heads. If you find bigger trouble than the odd pin-hole, a re-heat job is called for.

So on to the stays themselves, and there are a number of ways of tackling these little devils. The professional builders and experienced amateurs will prefer to silver solder plain rivets for the stays as being quick and effective. The main argument here is that once soft solder has been used on a boiler, no more silver solder can be used, and this constitutes something of a Rubicon for boilersmiths. However, if the thing has been proved sound so far, why should you need to use silver solder again? Far easier, particularly for the beginner, to avoid yet another cook up to red heat, and use high melting point soft solder to seal the stay-heads. The recommended material here is something like Johnson-Matthey's 'Consol' which has a melting point of 296°C and contains 1.5 per cent silver as well as tin and lead. This has good creep resistance at elevated temperatures but will run freely over the job. A melting point of only 296°C may sound a bit risky to some newcomers, since the inside of the firebox will be subjected to some pretty fierce heat when the engine is working hard, but, as long as the boiler has enough water there is never any problem. The water conducts the heat away so quickly that the copper plates do not get much hotter than the temperature of the boiling water. An alternative is Johnson-Matthey's LM 15, which is stronger. Copper is quite often used for stays, but the best material is drawn bronze, threaded, say 4BA for a $3\frac{1}{2}$ in-gauge boiler. Drill the sides of the firebox, throatplate and backhead, as shown in your boiler drawings, and tap to suit the stays. You may find you need an extra long tap for this, but, such things are available from the model suppliers. Take care to form clean threads in the very soft copper, then file off any burrs inside and out, and rub the metal bright all round the stay holes.

The stays themselves will need to be 'mass produced' in the lathe as follows. For 4BA stays use $\frac{3}{16}$ in rod and turn down a suitable length to pass through the firebox outer and inner wrapper and leave enough over to put a nut on the inside. Thread this 4BA using the tailstock die-holder and then start to part off leaving a head, say $\frac{3}{32}$ in thick, on the stay, but only cut half way through the rod. Take the rod out of the lathe and use it to screw the stay fully home. Then break off at the parting groove and go on to the next stay. Continue until all your stays are home, when you can file all the heads nice and smooth. Screw brass nuts on to the inside of each stay, or bronze if you prefer, and we are ready to sweat the lot tight with soft solder.

Pickle the boiler again for a few minutes, then rinse and dry thoroughly before painting liberally over the stay-heads with soft solder flux, such as 'Baker's Fluid' or Johnson Matthey's 'Soft Solder Flux No 2'. Warm up the boiler until the soft solder begins to run, then work it over all the stay-heads and nuts making sure that every one receives its due share. Allow the boiler to cool for a while, then wash under running water to remove the remains of the soft solder flux. Do not pickle the thing again because the soft solder will go black and horrible if you do. Do not throw away the pickle yet though, because it is handy for the boiler fittings. Before applying soft solder to the stay-heads it is worth checking on a couple of further items which might benefit from caulking over. First there are the support brackets which carry the weight of the back of the boiler on to the main frames. These are usually screwed on to the sides of the firebox and it is well worth fitting them now and soldering over the screws. Similarly, the hinge and latch arrangements for the firehole door are often screwed to the backhead and can be caulked at this stage.

If your boiler has longitudinal stays they will probably be of the threaded nipple type, in which case they should be fitted at the same time as the side stays and preferably sweated over too. I am none to keen on the idea of screwing the nipple on to the longitudinal stay at the same time as it is screwed into the smokebox tubeplate, but it does seem to work and is quite a simple arrangement.

With the boiler now completed we are all ready for the big test under hydraulic pressure, all the way to twice working pressure. This sounds dramatic, but it is not really, and is very easy to do.

The best plan for the beginner is to take his boiler along to the local model engineering society for testing. The boiler will then be issued with a certificate to say that all is well and that the locomotive can be used under the club insurance policy. You will find that all model clubs will require a boiler certificate before you can run on their track, but that most accept certificates issued by other clubs. Most likely the test equipment will be ready to hand at the model club. It consists simply of a large, accurate pressure gauge reading to 300 psi or so and a hand pump, as fitted to locomotives for emergency use. The procedure is to plug all the bushes and other openings in the boiler, very carefully, then fill it *completely* with water and apply the hand pump to one of the bushes. A couple of pumps will bring the pressure to 50 psi, or so, when a preliminary checkover can be made. If no obvious leaks or ominous bulges appear, continue gently to twice working pressure, probably around 180 psi, but do not go higher. Only a few strokes of the pump will be necessary since water is as near as damn it incompressible, it is only the slight movement of the boiler plates which allows the boiler to accept any more water once full. Hopefully you will be able to hold this pressure for several minutes, without further pumping, whilst all the joints are carefully inspected. Do not mistake leaks in your plugs for leaks in the boiler, as the former are much more likely. If you have to give the odd pump stroke to maintain pressure during the examination, this is okay, but if you need to pump steadily there must be a leak somewhere. If you do find significant leakage this can probably be cured by further sweating over, or drilling, tapping and plugging. Once the test is completed, let the boiler down and empty it out properly, then leave it to dry out, prior to installation of the frames according to your drawings. These will, no doubt, show fixings which allow for the expansion of the boiler in service, usually at the firebox end.

Incidentally, most clubs and, for that matter, prudent individuals, require boilers to be retested every two years. In this case, $1\frac{1}{2}$ times working pressure is all that is required and the boiler can be left on the engine. Only the safety valves and pressure gauge need to be removed. Should a boiler fail under hydraulic test then there is no hazard, other than wet feet, provided that it contains no air. People have, in the past, tested boilers to destruction, out of curiosity, and standard designs have gone beyond 800 psi, indicating that model boilers have a safety factor of around ten, which is reassuring. In point of fact, model boiler accidents are happily very rare, but it behoves us to make sure they stay that way.

Boiler fittings

Various bits and pieces are fitted on to the boiler of a locomotive to do a variety of vital jobs on the engine. Models designed for beginners will have fewer and simpler components than a scale locomotive, but even so there are a fair few items to manufacture and, once again, the cheque book may provide the quickest route to completion of the project. Many model suppliers produce fittings such as these, indeed, some specialize entirely in them and prices are not unreasonable. Complete sets of fittings can be bought for popular engines, such as 'Tich' and 'Rob-Roy', for sums in the region of £50 to £100, saving quite a lot of work. However, making boiler fittings is mostly fairly simple, pleasurable work and a welcome relief from the sweat of boilersmithing, so why not have a go yourself?

Brass is alright for these items, but gunmetal or drawn bronze is the best, if more expensive. A good plan is to visit the nearest large stockist of non-ferrous metals and ask to rummage through their scrap bin. My local people carefully separate bronze into a separate bin and sell this by the kilogram at half the price of stock bar. Their offcuts are ideal for the small parts we need to make. Components such as valve spindles will need to be stainless steel and the same tactic applies, though so little is needed that the model suppliers may be the best bet. Be prepared to do some small scale silver soldering here too. By and large it is simply a matter of working to the drawings of your design, but I will make some comment on each fitting and its function.

Safety valves

These are fairly easy, but be careful to D-bit the ball seat cleanly and ream the bore truly round, or the valves will tend to hiss continuously, which is annoying. It is possible to wind the springs yourself though a lot of people poke about in boxes of oddments until they find something apparently suitable. Make sure it is stainless though, or the safety valve could rust solid between runs with unpleasant consequences. If in doubt, buy from the model suppliers for a few pence. The blow-off pressure of the valves can be set during the first steaming session. They are first of all backed off to blow at a very low pressure, and then brought down using a suitable tool and a well-gloved hand, so that they release at the correct pressure.

Water gauges

It is extremely important for the driver to know the level of water in a boiler in steam. Too much water and the locomotive will 'prime', that is, take water into the cylinders, too little and the firebox crown will dry out, overheat, and ultimately fail catastrophically. The water gauge consists of a strong glass tube connected to the boiler, thus showing the water level. Many locomotives have two water gauges as a safety measure, and most gauges have a blowdown valve to get rid of bubbles which appear in the sight glass at times. From the construction point of view the important thing is to get the top and bottom fittings in line, otherwise the glass sight tube will crack when the seals are tightened. Use a piece of silver steel rod to check alignment. One problem with boiler fittings is getting them screwed in steam tight, but pointing in the right direction. The solution is quite simple — screw in the fitting to where you want it, but leave a gap of 10 or 15 thou between the boiler bush and the fitting flange. Measure this gap with feeler gauges, then part off a copper washer from bar, a few thou thicker than the gap. Anneal the washer and rub carefully with a figure of eight motion on medium emery paper, both sides. Try the washer with the fittings, and rub down its thickness until the fitting screws down tight, in the correct orientation. Do not use aluminium as this tends to corrode badly in service. Gauge glass can be obtained from the model suppliers as can suitably soft 0-rings to seal each end. Little more than finger tightness will seal the glass steam-tight, you will be surprised. Use a smear of plumber's jointing compound, a sort of runny putty, on the threads and washer when finally assembling the boiler fittings or, alternatively, use PTFE tape.

Clack valves
Non-return check valves, or 'clack' valves are fitted close to the boiler on all the various water inputs to the boiler. These prevent the back-flow of water to the pumps or injectors and are simple enough items to make. As with safety valves, take care with the ball seating to prevent leakage. Clack valves often have square plugs in the top and these make an interesting little exercise in filing, or, if you like, milling.

Blower valve
The blower valve is usually fitted to the backhead end of a hollow stay running through the boiler. Steam is collected via the turret and led through to the smokebox jet, via the blower valve, which controls the flow. It is usually just a simple screw-down device, but you might like to think about the handwheel because this gets very hot in steam. Articles appear from time to time in the model magazines with new designs for scalish, cool, handwheels, but the simplest plan is to make them from something like Tufnol. Turn up several of these and use them on any other wheel valves, such as the pump by-pass.

Steam turret
As mentioned, the steam turret, or manifold, provides dry steam for the blower and any other steam operated services such as injectors, whistle, pressure gauge, brakes or turbo-generator! No great problems here but the whistle operating lever may call for some fiddly file work. Some extremely able folks make their own pressure gauges, but this is very definitely something for the beginner to buy. Note the U-bend in the pipe leading to the pressure gauge as this is quite important. It fills with condensation at first steaming, and thereafter, protects the delicate bourdon tube in the gauge from the effects of hot steam.

Whistles
Working whistles are vital additions to the fun-factor of model locomotives.

Unfortunately they have to be very much larger than scale size to work, but they can be hidden away underneath the locomotive with no problem. They are best sited close to the firebox to keep them warm, otherwise the steam tends to condense before reaching the whistle, leading to some amusing, but not wholly desirable sound effects. Whistles can be bought, or quite easily made from drawings, though my first engine carries a Girl Guide whistle. With 80 lb of steam behind it Girl Guides are summoned up from miles around!

By-pass valve

The function of this valve is to control the amount of water pushed into the boiler by the axle pump. The output from the pump is divided into two, one line going to the boiler, via a clack valve, the other going back to the water tanks, via the by-pass valve. With the valve closed, all the pump output goes to the boiler. With it fully open, all the output returns to the tanks. By carefully cracking the valve, it is possible to set up a steady feed at the correct sort of rate for the conditions. Once again, it is a simple screw down device.

Blowdown valves

Boilers, like kettles, accumulate scaly deposits, particularly in hard water areas and most of these descend to the foundation ring area. This is the reason why it is known as the mud-ring in traction-engine parlance. One answer to the problem is to provide one or more blowdown valves just above the foundation ring. When the day's run is over, the fire is dropped out and the blowdown valve opened so that the contents of the boiler are driven out under remaining pressure, taking the scale deposits along with them. Once empty, the hot boiler will quickly dry out completely, thus protecting it from internal corrosion during storage. Some blowdown valves are simple screw devices, but the best incorporate a ball which revolves when the valve is open. This reduces damage to the valve by the fast-moving scale particles.

Snifting valve

Although arguably not a boiler fitting, this is probably as good a point as any to mention the superheater vacuum release valve or 'snifter' as it is often known. When the regulator is closed on a coasting engine, the cylinders act as pumps, evacuating the steam system downstream from the regulator. This means that there is nothing (or nearly nothing!) in the superheater elements, with the result that they will tend to overheat, burn, and eventually, fail. The solution is simply to provide a little ball valve which admits air to the wet header (see later) when the pressure in the steam system falls below atmospheric. It may be mounted in various ways on the smokebox; your drawings will indicate where.

Regulator

The job of this device is to control the amount of steam fed from the boiler to the cylinders, and hence the speed of the locomotive. A number of different designs were used in full size, most mounted in, or under, the dome, some in the smokebox, and similar devices have been used in models. Regulators in models however, need not be complex devices, all that is required is a screwdown device,

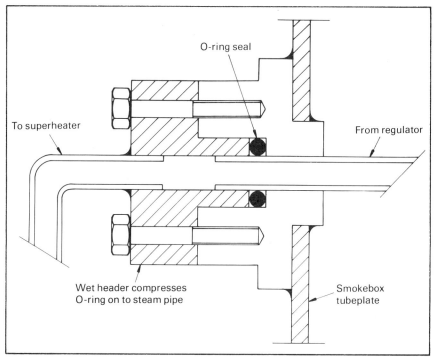

Figure 8.6: Section of the steam header and smokebox tubeplate to show one possible method of construction.

rather like a large size blower valve, as fitted on 'Rob-Roy'. If the spindle is threaded with something coarse, such as a ⅜ in or ½ in Whitworth thread, then the regulator will pass as much steam as required, with only half a turn from closed. See Figure 1.7 in Chapter 1. Construction of a screwdown type regulator is a matter of some pretty easy turning and threading, but assembly into the boiler can be a bit of a fiddle. Your chosen design will no doubt suggest a method of assembly but perhaps the best design employs an 0-ring for sealing the wet header to the steam pipe from the regulator. See Figure 8.6.

The regulator and steam pipe are inserted from the backhead and the steam pipe is eased through the flange in the smokebox tubeplate. It is then sealed steam tight by bolting down the wet header which compresses a suitable 0-ring on to the steam pipe. At the cab end the regulator is bolted to its flange on the backhead, and, finally, the steam collection pipe is screwed into the regulator body via the dome hole. Use PTFE tape to seal the little pipe and hold it in place.

Superheater

The purpose of this system is to further heat the saturated steam supplied from the regulator, before delivery to the cylinders. This imparts more energy to a given mass of steam and thus produces a considerable saving in water consumption, but it does not greatly increase the power of the engine. True there may be some reduction in coal consumption, and condensation in the cylinders will likely be eliminated, but the only real benefit for models is the improvement in water consumption.

Superheating is achieved by leading the steam pipe through special large flues in the boiler, so that further heat can be imparted to the steam. Older designs tended to be of the 'spearhead' type in which the return bend of each element was within the flue tube, but more recently the radiant type of superheating has been introduced.

This reached right into the firebox and is hence much more effective. Stainless steel tube is the material for this job, and the return bend, often made using a suitable drilled block, will need to be welded, or at least brazed, to withstand the heat. There are often several elements in a superheater, connected in series, with one end fitted to the 'wet header' and the other unioned on to the cylinder steam pipe.

Grate and ashpan

These items are not generally considered as boiler fittings, but they are none the less part of the boiler system and merit a word or two here. Cast iron grates are available for most designs of locomotive and these are cheap and effective, but they do not last very long. A season's good running will probably burn out an iron grate, but stainless steel will go on for several seasons.

You can either make the grate up from the stainless bars, or use one of the ready made types as advertised in the model press these days. Ashpans are generally bent up from mild steel sheet and brazed as necessary, according to your drawings. They can be more or less complex, depending on the proximity of axles, frames and so forth. One very desirable feature of any grate and ashpan assembly however, is that the two should be readily removable from the locomotive in an emergency, such as gauge glass failure. The usual scheme is to retain the grate and ashpan assembly under the firebox by use of a pin through the main frames. In an emergency, and also at the end of the day's run, the pin can be withdrawn and the grate and ashpan dropped clear of the engine completely. This is going to be more difficult with locomotives in which the firebox is mounted above an axle, either coupled or trailing, but if you can arrange it, a quickly removable fire will make for greater peace of mind.

Injectors

Again, these are not exactly boiler fittings, as such, but here is as good a place as any to mention them. These amazingly cunning little devices feed water into the boiler using steam from the boiler to do so. This sounds rather like something for nothing but it is all fairly simple science really, with the help of friend Bernoulli from Chapter 1.

Essentially, an injector consists of three specially shaped cones, a steam cone, a

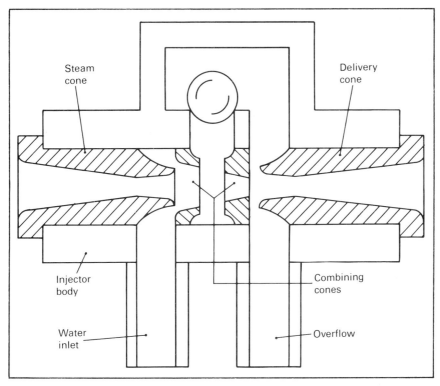

Figure 8.7: Schematic section of an injector to show the relationship of the various components.

combining cone and a delivery cone. The function of the steam cone is to take steam from the boiler, via the turret, and expand it in a sort of jet. The cone is so shaped that the jet accelerates to a high velocity, whilst its pressure drops to atmospheric or even less. This high velocity steam is injected into a small reservoir of cold water from the tender or tanks of the locomotive, and on into the combining cone. Here the energy of the jet is handed over to the feed water, and the resulting fast-moving stream is then directed to the delivery cone, where Bernoulli does his stuff again and velocity is swapped back to pressure. Indeed, the pressure is now high enough to enter the boiler and so water is thus fed to it. Naturally, the simple swapping of pressure to velocity and back again could not provide the energy to push water into the boiler, indeed energy is lost in the conversions by friction. The answer is that the necessary energy is supplied by the heat of the steam which is condensed into the feedwater. This was all made to work in full size during the last century, although model injectors were a little later in coming!

The advantage of injectors is that they will supply the locomotive when stationary, and have no moving parts to wear out. They are rather prone to blockage however, particularly in the smaller sizes, and need cold feedwater to operate properly. This latter point is a problem on tank engines where the boiler tends to heat the water. Injectors are usually mounted low down on the outside of the engine, close by the cab. In this position, water can flow into them by gravity from the tanks or tender, and the body can be kept relatively cool. They are easily accessible for dismantling and cleaning and you can also see if the overflow is running, indicating that the injector has not 'picked up' properly. The procedure for starting up an injector is to open the water valve, and then the steam valve. A high-pitched whistle, or whine, means that the device is working, a flood of water from the overflow means that it is not and the valves should be turned off for another try. Once running, injectors will continue to feed despite changes in boiler pressure.

Once again you can buy injectors quite cheaply, or, if you would like to try a little careful turning and fitting, then have a go at making one yourself. Several of the more advanced locomotive books will tell you about injectors but LBSC's *Shop, Shed and Road* has a full account of the construction of these clever pieces of locomotive equipment.

This just about wraps up boiler fittings for the beginner so that we can now go on to the final area of construction; the superstructure.

Chapter 9

The top bits

Smokebox assembly

Smokebox assemblies come in several forms, but the most common arrangement on locos aimed at beginners is a tubular smokebox, resting on a cast structure called the saddle.

The smokebox can be rolled from plate but the easiest plan is to use a suitable length of solid drawn tube, carefully chosen to be a push fit on to the boiler barrel. Steel is a cheap material for the job, but brass or copper will get over the corrosion problem and are, perhaps, a little easier to work. To clean up the tube to length you can use a carrier as described for the boiler barrel or you can get away with a single wooden disc pushed into one end, allowing you to grip the tube in the three-jaw. Use the lathe tool to scribe a faint line along the tube before removing it from the chuck, then you can centre punch the place for the chimney along this line.

The hole for the blast pipe needs to be marked out exactly opposite, and the way to do this is to wrap a tape measure around the smokebox and measure its circumference, then measure precisely half this value round from the chimney line to give the position of the blast pipe line. Builders of smaller engines will probably get away with drilling the holes for chimney and blast pipe directly, using the wooden disc to support the rather flexible smokebox tube. With larger engines, a better plan may be to bolt the tube to the lathe saddle at the correct height and bore the holes with a boring bar.

Actually, the hole on the top of the smokebox is not strictly for the chimney, but for the petticoat pipe. This item is made from soft copper tube, belled out at the lower end and silver-soldered to a square plate through which it passes. This plate is curved to be a snug fit on the inside of the smokebox, and held to it by four countersunk screws. The chimney itself is made a push fit on to the petticoat pipe, and its base covers the four screw heads; cunning stuff, eh? Take some care with this assembly, or the chimney will tilt, which looks terrible on long chimneyed locos.

You will also need a hole in the bottom of the smokebox for the steam pipe. The saddle is often, as mentioned, a casting, although it may fairly easily be fabricated from plate and silver-soldered together. It fits snugly between the main frames and the smokebox sits on top of a suitably curved surface. If the casting is a nice smooth, accurate one, then a clean up with files may be all that is required. If not, then some machining will be called for. The flat faces can be dealt with by an end-mill, but the curved surface for the smokebox presents a more interesting machining problem. The way to tackle this one is by the process of fly-cutting. This involves a boring bar held between the lathe centres, carrying a long cutting tool set so that the distance from its tip to the lathe axis is the same as the radius of the smokebox. The saddle casting is then clamped to the boring table at the correct height, and then traversed slowly under the revolving cutter. Go pretty slowly to begin with because the cut is intermittent and a hard spot in the casting could cause problems. The same arrangement is used to tackle the base of the funnel and, with a slight change in radius, the base of the dome. In some cases you will find that the lathe cannot accommodate the full revolution of the fly-cutter, in which case you may have to resort to hand turning of the cutter. The tool is pulled through the cut and then turned back and the job advanced a few thou, then pulled through again and so on. This is like a shaping machine, but working on a curved surface. It may be a little laborious, but it is very effective.

The front of the smokebox is usually a turned ring, very often a casting, finished

Figure 9.1: Cross-section of a typical locomotive smokebox to show the steam pipe arrangements.

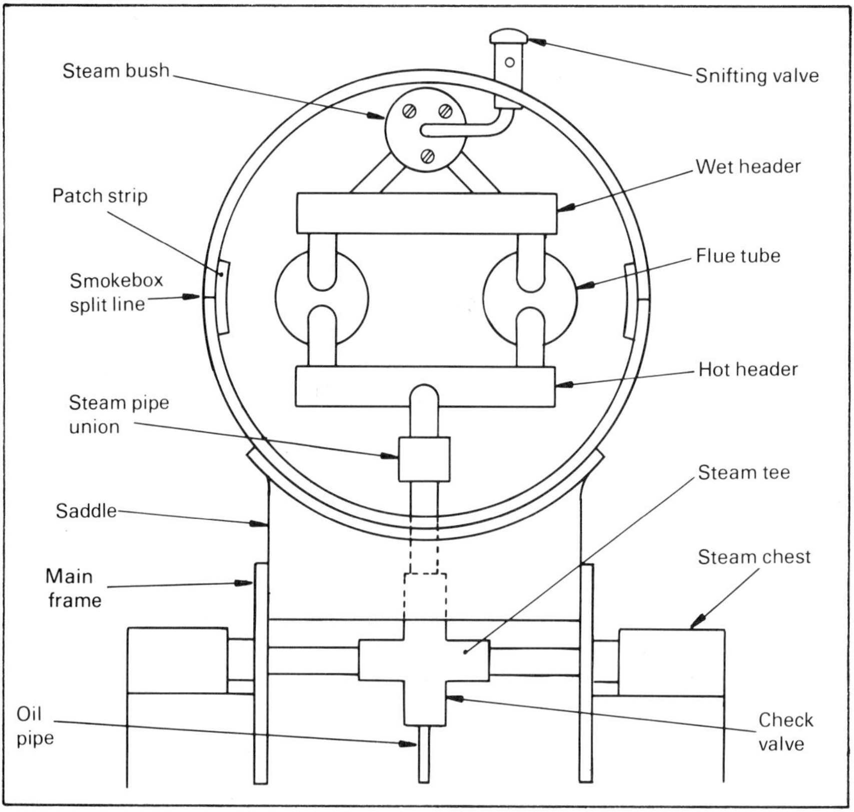

to a tight push-fit into the smokebox tube. The front face carries the smokebox door, again often a casting, the hinges and so forth. The door hinges can be a bit fiddling but take care to get them level and properly spaced for the sake of appearance. The door should close snugly against the front ring so as to be reasonably air tight, thus maintaining the vital smokebox vacuum.

One problem, however, with the conventional smokebox, particularly in smaller locos, is the accessibility. There are quite a few pipes to connect up in this area during assembly, and during running the whole lot gets plastered in soot. This means that alterations and adjustments are rather tricky, if not impossible. One answer is to make the front ring easily removable which improves accessibility, but the best bet is a split smokebox, with the top half removable.

Figure 9.2: Cross-section of the same smokebox showing the exhaust pipe arrangements. Bear in mind that this plumbing must be combined with that in the previous diagram and that all must be accommodated within the confines of the smokebox. This can make the extra work of fitting a split smokebox well worthwhile.

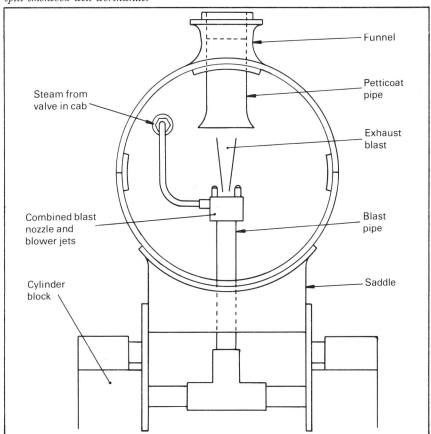

This can easily be arranged with a backing strip along the split-line, soldered to the bottom half of the smokebox. A row of countersunk screws tapped into the backing strip then retains the top half, reasonably unobtrusively.

Next comes the funnel or chimney which, once again, will probably be available as a casting, usually in gunmetal. First operation is to bore the casting to a nice push-fit on the petticoat pipe, and to chuck the job a simple split bush can be used. Turn a length of wood to something a bit bigger than the funnel rim, bore to the diameter of the funnel body, then unchuck and split it endways. The bush can then be placed over the funnel body and the whole lot gripped in the three-jaw for boring the casting. If your design has too short a chimney to accommodate a bush, or one too big to go in the three-jaw chuck, you will have to devise something else! Having bored the job, push it on to a suitable mandrel and turn the outside and rim to drawing. The base cannot be turned because of its irregular shape so you will have to brush up your filing skills on this one. If yours is the sort of engine to carry a bright funnel cap, turn it to a fine finish and then polish it in the lathe. Cover the lathe bed with a rag to stop polishing compound getting into the bed ways. You might also tackle the dome casting at this stage because it will need very similar treatment. The main difference with the dome is that there will probably be a chucking spigot cast on the top, so there will be no need for a split bush.

The final operation on the chimney is to finish the curved bottom face to fit snugly on to the smokebox. Careful filing may be all that is needed here, or as mentioned earlier for the saddle casting, this face can be fly cut. My photographs show one set up for doing this job on a Myford lathe, using a simple homemade fly cutter. The only tricky part is getting the casting lined up truly at centre height. Go very gently, advancing (or in the case of my set-up, retracting) the job only a couple of thou at a time and keep the tool sharp with frequent stonings.

Having made all the smokebox components we are ready to mount the boiler and fit up the smokebox plumbing. The boiler is pushed firmly into the smokebox whilst it is supported at the firebox end on the brackets fitted at each side. These rest on suitable rails fixed to the frames, with clips to keep the brackets down on the rails. This set-up allows the boiler to expand whilst in steam.

The superheater elements can be pushed into the flue tubes and the wet header end connected to the tubeplate bush, whilst the delivery end is unioned with the steam pipe, as mentioned in Chapter 7. The snifting valve, if you have one, will have similar, but smaller, connections. The blast cap can be screwed on to the blast pipe, and the blower plumbed into the hollow stay through the boiler. At this stage check carefully that the blast nozzle points directly up the funnel and is centred on

Top left *There will be some interesting pipework in the smokebox of this three-cylinder 'Pacific' loco! Note the double chimney.*
Centre left *Castings for the saddle and funnel of 'Rob-Roy'.*
Left *Machining a 'Rob-Roy' funnel on a mandrel. The curved flare of the rim can be shaped with a hand graver or, as I did, by judicious twiddling of the cross-slide and top-slide at the same time! The curve at the base can only be tackled by hand filing. The emery stick is for polishing the rim; note the rag to protect the lathe from the abrasive dust.*

Top *Fly cutting the base of the funnel with a home-made cutter. The only tricky bit here is to get the funnel set up truly at centre height and at right angles to the lathe axis.*

Above *The finished funnel, a satisfying piece of lathe work.*

the petticoat pipe. Insert a piece of suitable silver steel into the blast nozzle to do this test. If the alignment is anything but perfect, then do some judicious bending or fitting until it does. It is worth spending some time getting things neat and tidy in this area because it is rotten work trying to make alterations once the engine has been in steam.

One problem here is the sealing of the steam and exhaust pipes as they pass through the bottom of the smokebox. In theory you can ream nice neat holes for these to pass through, but in practice this rarely seems to work out and things have to be 'eased' to fit. This leaves you with some awkward gaps to fill with something resistant to the heat of the flue gases. In the old days asbestos was recommended, mixed to a putty with water and pressed into place to set hard later on. Asbestos is now rather frowned upon since it produces tiny airborne fibres which become deposited in the lungs and cause serious problems. As an alternative try some of the rock wool, as suggested for boiler building, or even fire cement.

Running boards, tanks and tenders

The next items after mounting the smokebox and boiler are usually the running boards and valances, with the splashers, if your design has them. Running boards are very often simple flat plates resting on the buffer beams and the tops of the main frames. They will have various cutaways to accommodate pipework, reverser stands, and so forth. Brass is recommended in the smaller gauges, although steel is perfectly adequate, and a lot cheaper. You will notice, no doubt, from your drawings, that the running boards overhang the buffer beams and valances by a fraction of an inch, around about the thickness of the plate used. The same thing is found on bits like the tops of side tanks and is quite an important feature in getting the 'looks' of the engine right. Flush finished platework is non-scale practice and looks terrible. Take care to make the edges straight and sharp. De-burr them, but do not round off the arrises. The valances are usually made from suitable angle, riveted or soldered to the outer edge of the running board. The curly decorative bits often fitted to blend into the line of the buffer beam can be made from plate and soldered in place.

Mention of riveting the valances calls for a little further comment, applicable to all the platework on the engine. Some folk develop 'rivet fever' at this stage in building a loco, and put row upon row of snap-head rivets, frequently oversize for scale, along every seam in the platework. This may look impressive to them, but to the cognoscenti it looks pretty bad. It is true that many full-sized locomotives had riveted seams in some places, such as on tender tanks and around the smokebox, but round-head rivets were not evident all over the engines, particularly the more modern ones. If yours is a scale or near-scale design then find some pictures of the original, or better still, if a preserved example exists, climb about on that, and look at the rivet detail. On a free-lance engine like the ubiquitous 'Tich', soft solder the valances or flush rivet. The latter process is quite easy, just provide a countersink on the top face of the plate, using a drill about twice the diameter of the rivet being used. Push a soft copper snap-head (round-head) rivet through from below and snip off about one rivet diameter above the surface of the plate. Now hammer the rivet gradually into the countersink using a light ball-pein hammer whilst supporting the head of the rivet on a suitable rivet-snap, to preserve its figure. Aim to fill the countersink with copper and leave a few thou proud, then clean off by rubbing carefully with a smoothish flat file. This will result in totally invisible rivets, but quite a strong joint. The valances, water tanks and cab of *Lady Caroline* were flush riveted by a complete beginner, so have a look at the photos of this engine for an example.

If your project is a tank engine, particularly a side-tank job, then the next episode will likely involve making up the tanks. A silver medal-standard loco builder in my local engineering society always uses galvanized steel for his tank and tenders, but for the beginner brass is the only sensible material, the half-hard,

sheet variety. It is depressingly expensive, particularly if you have opted for a large engine, but it is well worth the money. There are no corrosion problems and brass is nice and easy to work. Practically every article one reads on model locomotive construction, right from the 1920s and '30s up to now, contains the phrase 'the high cost of materials these days ...' or something similar; actually, the cost of materials has fallen steadily in relation to average earnings and although you may spend several hundred pounds in the end, it will be nicely spread over perhaps two years of construction. This is equivalent to two or three pounds a week, the cost of forty cigarettes or three pints of beer at the time of writing. Seems like good value to me.

Your drawings will no doubt suggest a method for putting the tanks together

Above *This tender is for the 3½ in-gauge '8F' seen earlier on. Super work from a beginner!*
Above left *We have seen this 'Rob-Roy' earlier in the book. Now it has a boiler and platework and is not far from steaming.*
Left *Here is a nice tender in 5 in-gauge with some detail work.*

with brass angle and copper rivets. Take care with the brass sheet not to put unwanted bends into it, or the tanks will look as if they are made from warped wood. When it comes to putting 90° bends in half-hard sheet, do not anneal it, because it will never be flat again afterwards. Instead, clamp the sheet between stout pieces of wood at the bend line, and hammer the top over using another piece of wood as a drift. Better still, see if you can borrow a sheet metal folding machine which will make platework sheer bliss. Having access to tools such as these can speed the job along splendidly.

Having made the tanks, they can be caulked inside with soft solder. Swill the inside around with suitable flux, such as Johnson-Matthey's or the famous 'Baker's Fluid', then warm up and run solder all over the inside. Cool and wash thoroughly to remove any traces of flux which may set up corrosion later on. Here again, the beginner might choose the modern alternative and bond the tank components together with one of the new 'Loctite' products. These are beyond reproach in terms of strength, and much quicker than riveting.

Now is the time to fit the hand-pump into one of the tanks, if this is its designed position. You can buy this item ready-made, or make one up like the axle pump.

It is a good plan to make the tank tops removable, so retain these with some small countersunk screws. Builders of saddle tanks will have to contend with slightly different problems, and will need a set of bending rolls, rather than a folding machine.

If you are making a tender engine, then the tender body is very similar to a large side tank, whilst the chassis is a simplified version of the loco chassis. Once again brass is the material for the tender water tank, and your drawings will no doubt indicate the way in which the thing should be put together.

Plumbing, lagging and cabs

By now the locomotive has gained most of the bits and pieces it needs to run, but is

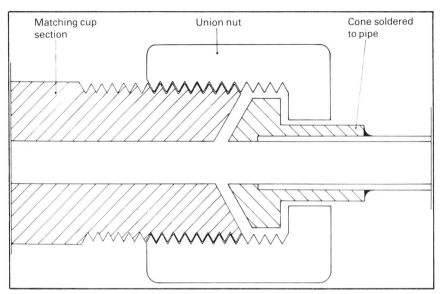

Matching cup section   Union nut   Cone soldered to pipe

Figure 9.3: Cross-section of a pipe union to show construction.

bare of cab, boiler lagging and so forth. This is then the time to tackle the plumbing, whilst the works are still accessible. This means making up the pipe-work and connections between the various boiler fittings, pumps, tanks, tender and so on.

The actual pipe runs are made in copper tubing of appropriate size; $\frac{1}{16}$ in bore for oil pipes, $\frac{1}{4}$ in bore for water feed to the axle pump, and so on. One way to achieve neat pipework is to bend up a dummy from thick copper wire to begin with, and then use this as a model to form the pipe itself. Anneal the tube by heating to red and quenching first, and pickle it to clean off the scale. The tube can then be generally bent using just your fingers, without forming kinks. Thicker wall tubing is easier to bend without kinking, but you may need to bend part way, and then re-anneal before completing the bend to pattern. Joints between the pipe-work and the fittings are made by classical unions, that is, a cone is soldered on to the pipe, the fitting is provided with a matching cup, and the two are held together by a union nut (see diagram). You can buy these bits and pieces quite cheaply, or make them very easily on the lathe. Commercial cones, or nipples, are made in brass, they can be made in copper if you wish, but in either case they should be silver-soldered on to the pipe. Remember to put the union nut over the pipe before soldering, because in many cases, you cannot put it on afterwards! This type of joint can be made down to very small sizes and is pressure tight with only a very moderate torque on the union nut, little more than finger tightness, so do not overtighten it. It is worth taking care to make a neat job of the plumbing, particularly around the backhead because, like paintwork, a tatty job here can spoil an otherwise fine locomotive.

Full size locomotives had lagging around their boilers to reduce heat loss by radiation, and small locos can be treated similarly. The thermal benefits of lagging are probably not enormous but other benefits are gained. For one thing, the boiler is cooler to the touch so you are less likely to be sued by the parents of over-inquisitive children. For another, all those blobs and runs of solder on the boiler are covered over; and last but not least, the cleading sheet provides something to fit things on to, like handrail knobs, without penetrating the boiler. Full size engines and earlier models were lagged with asbestos which probably killed a good many unwitting labourers. Model builders can use glass fibre, but about the simplest material is balsa wood. This can be bought in suitable thickness sheet and wrapped neatly around the boiler. It is easy to make the cut-outs for dome, check valves, safety valves and so forth, and the whole thing can practically be made in one piece. This will then stay in place whilst the cleading sheet is fitted, unlike fibrous materials which are difficult to deal with.

The cleading sheet is the thin metal plate wrapped around the boiler, outside the lagging, and usually held in place by a number of prominent bands around the barrel. Thin brass sheet of about 26 swg is the material for the job, but make up a cardboard dummy before hacking up expensive brass. The model suppliers can provide brass strip suitable for boiler bands and this is usually joined by a small nut and bolt passing through pieces of angle soldered to the ends of the band. On some tank engines you will need to fit the boiler bands before finally putting on the

Inspiration for the new-comer; the cab of an exhibition standard 5 in-gauge Great Western 'Manor'.

tanks. A further advantage of boiler lagging, or cleading as it is often known to enginemen, is that the outer sheet can be properly painted, away from the engine. The same is true of the boiler bands, which were decoratively lined out in many full size engines.

The next job, and probably the last major one, is the cab structure. Cabs come in all shapes and sizes, from the simple weather board found on small shunting engines, to the elaborate shapes fitted to more modern express locos. Generally speaking, their construction is similar to the rest of the platework. Most people will go for brass, especially for smaller engines, but steel is perfectly okay. Aluminium alloy would do at a pinch, and is very easy to work, but is going to be prone to damage over the years. One advantage of steel is that it generally takes paint better than brass, but more of that later.

Windows, particularly those facing forwards, will look better for glazing, although this can be a little fiddly. Perspex is nice and easy to work, but will go dull and scratchy pretty quickly. Glass is ideal, but very difficult to work unless you can find some microscope cover slides of the diameter to suit round windows, or you could try asking a watchmaker for help. Sheet mica is another alternative if you can lay hands on some.

When producing the front face of the cab, cut a cardboard dummy first, as with the cleading sheet, so as to avoid wasting precious metal on a botched job. Take care that your cab does not make it impossible to reach all the controls to drive the engine. If necessary, make cut-outs or removable sections so that it is possible to fire the engine on the run without the aid of tweezers. This mainly applies to tank engines, but the roofs of many tender engines can make life difficult for the driver if not removable in some way.

Take care with the cab to get straight edges straight, flat surfaces flat and curved surfaces even. As with running boards, edges should be de-burred but not rounded. Many designs call for a beading around the edge of cab platework, and this is no problem. The model suppliers produce half-round brass wire for the job. Bend and hold in place with toolmaker's clamps and then soft solder. On my first engine I tried to play clever and used a row of paper clips to hold the beading wire in place for soldering. One pass with the butane torch to warm the job and the paper clips immediately lost their temper and fell off! Luckily, the only harm done was to my ego.

At this stage, the model loco is very largely complete, certainly to the level at which it can be trial steamed. Indeed it can be run without cab, lagging and so forth, but better to take it nearer to completion. The other extreme is to carry right on to the end, with detail work, paint and polish, but this is a trifle risky for the beginner.

The best place for a first trial is not the club track on a crowded Sunday afternoon, however confident you may feel! Far better either to run at the track when nobody else is around, or at home on the bench where adjustments, alterations or, God forbid, repairs can be made well away from public gaze. If you can invite an experienced and sympathetic loco man to come and look over your shoulder, then do so. He should provide moral encouragement if problems emerge, and praise if they do not. Raise the loco on blocks for the test run, or better, make up a simple carrying frame and run on that. Steam raising can generate quite a bit of smoke, and the first few turns of the wheels will throw hot oily water all over the place, so the domestic kitchen is best avoided for this operation! I plan to describe steam raising and driving in the next chapter, so let us assume that all has been settled and that the engine runs satisfactorily. We can then proceed to the next stage, that of detail work.

Detail work

Fitting the detail work on to an engine can be a time consuming business, but is highly pleasurable. It is the addition of all the small bits and bobs that turn a glorified toy into a fine model and, incidentally, into a valuable one, for those who think in terms of hard currency. Little heavy work in terms of machining and fitting is involved, and there is little in the way of commitment; that is, if the piece goes wrong, it can be junked and another started. The working engine is not affected and can be run from time to time as the extra parts are built on. Detail parts can mostly be made from odd ends of scrap which means you can now start thinking about spending money on castings for the next project! Beware, though — do not start on the next one until this one is totally, one hundred per cent complete. The world aready has ample stocks of uncompleted model steam locos hidden away in dusty corners.

Detail work can be taken as far as you fancy. LBSC was not a great one for scale intricacies, preferring practical, working engines to glass case exhibits. He used to fit buffers, hooks and handrails, but not much else. Going to the opposite extreme, there are those who go to enormous lengths to duplicate the features of the original locomotive and win well-deserved medals at the exhibitions as a result. It is a case of horses for courses, and you must decide what is appropriate. There is little point putting the full works, sanding gear and turbo-generator, etc, on to a free-lance shunter, but if you have made a good job of a near-scale loco, like 'Princess Marina', it is a pity not to add some realism to the final job.

Plenty of beautiful detail work on this 5 in-gauge 'A3' 'Pacific', the famous Flying Scotsman.

Working buffers are absolutely vital for the public image. Hosts of small children will delight in pushing on the buffers whilst their doting mothers say 'Look — even the little buffers work!' Such is the reward for hundreds of hours of work for the model builder. They are fairly easily turned up from bar or castings, but do not use cast iron for the heads, because it is hard to get a good finish and it then rusts. Use mild steel, put on a fine finish and keep it fairly oily in storage. Alternatively you could use stainless, with a good cutting compound to help the lathe, but its shine will give your model an unauthentic appearance.

It would be nice to ream all the stocks to take the heads, but the lead on commercial reamers means that you cannot get a parallel hole to the bottom. You could bore them all to the same cross-slide dial setting, or make up a D-bit for the job. The heads call for a bit of brain work, because they can be awkward shapes, with spherical faces and so on. Note that they usually have a generous radius at the join between head and stem which will call for a well rounded lathe tool. The mouth of the stock should be radiused to match, or the buffer will tend to jam in. By now, you should be able to devise a scheme for turning the buffer heads to a matching set with a nice sliding fit in their stocks. Tender buffers present a slight problem since the frames get in the way and there can be no extension behind the buffer beam. Your drawing will suggest something suitable, but avoid having to drill right through the buffer head to fit the central rod. This is unnecessary, use Loctite or something similar to retain the rod.

Left *And yet more detail; the back end of a 5 in-gauge 'Britannia'.*

Top right *A commercial scale coupling by Blackgates Engineering. Good quality and a reasonable price. The quick route to detail work, and why not?*

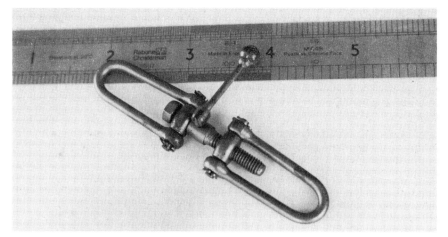

Coupling hooks and couplings are almost as important as buffers. Quite often one sees a good loco fitted with poorly made coupling hooks, roughly filed to the wrong profile, probably as a hasty afterthought. Mark out a suitable blank for the hook from your drawing, then cut out the profile carefully before radiusing the outline as per prototype. Simple shunting engines will just need three link chains by way of couplings, bent from mild steel or stainless steel rod of the appropriate diameter, and brazed together. A small jig may make it easier to bend up matching links. Larger type locos will look the better for carrying proper screw couplings. You can set about making these, but several model loco specialists offer them ready made at a very reasonable price, perhaps £5 each. However, scale couplings should only be used on the front of an engine intended to do serious passenger hauling. At the back, something much stronger and more positive will be needed. Best thing here is a pin passed through a double bracket and nutted or fitted with a security pin. A runaway locomotive is a pretty dangerous object, apart from the heart-breaking damage it will do to itself, given half a chance (see Chapter 10).

Brake gear is well worth fitting to any loco, but only for cosmetic purposes, because it will be completely ineffective for stopping even the smallest train. As in full size, the only real way to stop the train is to provide brakes on as many of the carriages as possible. The only practical use for brakes on the engine is to stop the thing from moving when parked on the steaming bay; however, they do improve the looks no end. Somehow, an engine looks quite naked without proper brake gear, and most designs incorporate the wherewithal to fit it. The gear may vary from a simple, screw-down column in the cab to steam operated, air operated, or even vacuum brakes. The average beginner will probably be quite content with the work involved in making hand brake gear. The only problematical components are likely to be the brake blocks, which need to be radiused to match the wheels, split at the back for the hanger, and generally profiled to look like the full size article. The best solution here is to use one of the cast rings of brake shoes produced by some of the model suppliers. The whole lot can then be radiused and slit by simple lathe turning operations.

Handrails are prominent on many, if not all locomotives, and these are easy enough to reproduce. Do not bother to make the stanchions, but buy them by the packet, ready-turned and chrome plated by the model supplier. They have threaded shanks and can be screwed direct into threaded holes or held on with nuts at the back. It is very easy to fit them to the cleading sheet, water tanks, smokebox, bunker, or whatever. Buy stainless steel rod to match the stanchions and make up the actual handrails from this. True, mild steel would be more accurate in terms of scale practice, but it does need some looking after.

Footsteps are almost as prominent as handrails, and can quite easily be fabricated from brass sheet and fitted each side of the cab. Not quite so easy is the distinctive 'chequer plate' used on full size loco steps and footplates, rolled from sheet steel. There are two solutions for the modeller. If impecunious, then get hold of some of the fine diamond-pattern expanded aluminium supplied for helping to patch up dented cars. It comes with the packs of resin filler found in car spares shops. Lay this down on the steps and footplate, apply a liberal coat of dull black paint, and the result looks exactly like chequer plate. For the more wealthy, you can buy specially produced miniature chequer plate from certain model suppliers, in small sheets. It makes a nice touch of detail on any engine.

Guard irons were fitted to all full size British locos to knock debris off the track ahead of the loco wheels. Model guard irons will no doubt do something similar for model locos, and improve detail appearance at the same time. Some had quite acute bends in them, and you may need to bend hot if using steel for the job.

Another important feature on big engines were nameplates, number plates, shed plates, makers' plates and so on. These can be reproduced in miniature by photo-etching, a process you can do yourself if you wish, but one that is best left to the professionals. These folk advertise in the model press and produce standard sets of plates for certain engines, or will produce plates to your specific requirements. The process is one of photo reduction, so you can save a lot of money by doing the 'artwork' yourself. This involves drawing out the plate as large as possible and carefully inking or painting on the letters. This artwork is then photographed and the negative projected on to brass plate coated with light-sensitive varnish. Where the light strikes the varnish (usually ultra violet light) it is softened, where protected by the film it remains hard, so that when the plate is dropped into an etching solution, such as ferric chloride, the areas exposed to light are etched away. The result looks much like the cast plates with raised letters so typical of the real engines. For simple nameplates one can use a poor man's version of the process, as was done for *Lady Caroline*. In this case, Letraset was used to form the letters on previously cleaned slips of brass. The back and edging were painted, and the plates were hung in concentrated ferric chloride for 15 minutes. The paint was then removed, and the fronts given a coat of gloss red. When this was fully hard, the plates were rubbed gently, face down, on fine emery paper to give the result you see in the photographs. Ferric chloride is used for etching printed circuit boards in the electronics industry, and can be bought from some electronics suppliers. Failing that, your friendly chemist might be able to help.

An interesting model this one. My picture shows the lubricator driven by the crosshead and a steam feed pump.

Oil lamps were carried on full size engines in set patterns to indicate the type of train they were hauling, indeed, pretty similar lamps are still hung on the back of modern diesel and electric trains. These lamps make a pleasant addition to both front and back of the model. Recourse to the cheque book is once again recommended here, because you can buy some sweet little die-cast lamp kits suitable for various gauges from the ubiquitous model suppliers, as with screw couplings.

Further detail work can be added according to your prototype and your inclination. Items like sanding gear are rather like brake gear in that they can be made to work in model form, but are largely cosmetic in value. Toolboxes, Wakefield lubricators and so forth can be added in dummy, solid block form. Those building foreign engines such as 'Virginia', LBSC's classic cowboy film locomotive, will have fun making up cowcatchers, spark-arresting funnels, huge headlamps, bells and so on. For those with a penchant for watchmaking it is possible to manufacture miniature turbo-generators, and hence light up the cab for night running! The list of details is practically endless, so set your sights before the job overwhelms you!

Painting and lining

There are two basic ways of going about painting a loco, and the builder will have to decide at the outset which course to take. Number one option is to paint up each part as you go along; number two option is to finish the loco completely in bare metal, have a few test runs, then strip down and paint the parts before reassembly. The latter option is more time consuming but, in the end, will produce much better results. The approach I myself have used is something of a compromise between these two extremes. This is to paint the main frames, stretchers, buffer beams and wheels at the assembly stage, but then to put the rest together without paint, and test run. The superstructure can then be fairly easily dismantled for painting, but the valve-gear and so forth can be left intact without disturbance. The finish on the frames may suffer during assembly of the engine, but this part is much less prominent than the boiler, cab, tender and so on.

If you decide to paint your loco after completion but stripped down, you are faced with some further choices; first of all whether to go to a professional or to do the job yourself. There are people who advertise in the model press, particularly in the table-top size range, specializing in the painting and lining out of model locomotives. No doubt one of these would produce a very fine finish, but probably at a cost of £400 or £500 on, say, a 5in-gauge tender engine. If you do decide to use a professional, though, ask to see some of his previous work and check that he is going to be worth his sodium chloride!

The next option, and for my money probably the best, is to take the parts which require painting to a commercial stove enameller. The ideal sort of company to look for is one which deals with painting the cases of electronics equipment, instruments and the like. They will know what is involved, and how to put on the finish you specify. If you can, get them genuinely interested in this little job, discuss it carefully with them and explain just exactly what you want done to each part. If the people are really not interested, go somewhere else because you risk loss or damage to your valuable loco parts in an apathetic environment. There is nothing to beat a properly stoved finish, but you may find yourself limited on colours. This is no problem if you want Henry Ford's famous black, but subtle railway colours may not be available in the sort of paint the stove enameller uses. There is at least one paint company which specializes in railway colours, advertising in the model magazines, but professionals may not wish to handle this type of paint; in which case it is something to tackle yourself.

The first item to deal with is probably the smokebox assembly, which gets hot in service and must be treated accordingly. These bits (smokebox, saddle, funnel and door) were invariably black in full size, so visit your local car accessory shop and buy an aerosol can of heat resisting matt black, as used to make cars go faster, or something. Not the graphite stuff, but genuine matt black finish. Rub the smokebox over with fine emery to provide a key, make sure that it is grease-free, and spray on a number of very thin coats. This paint dries very quickly and one need only wait a few minutes between coats. The parts can then be cured according to the instructions on the can, probably in the domestic oven.

The next job is the preparation of the rest of the parts to receive first primer and then top coat. Obviously all the parts should be rubbed down smooth with fine emery paper to provide flat surfaces with some 'key' for the paint to get a grip on. All the parts should also be absolutely, surefire grease-free, so give them a good bath in petrol or meths, out of doors. Steel parts can be given a coat of zinc primer if you wish, but brass or copper surfaces must be treated with etch primer first. Without this stage, paint will simply flake off cuprous metals in service. Etch primer can be obtained from the ever-faithful model suppliers, and is a rather thin, watery-looking varnish which can be sprayed, but seems to go on well enough with a brush.

Now at last, you can begin to think about the final finish, and think about it you should. Not only is there the question of colour, of which there is an endless variety, but there is also the question of dull or gloss finish. Most full size locos were certainly finished in gloss originally, but in service they quickly dulled down to a sort of oily sheen. To my mind, a glossy model looks cheap and toy-like, and for a sophisticated, refined result, I much prefer a dull finish. Not matt, mark you, but dull, or egg-shell as it has been called.

The best way to apply the top paint is with a proper high pressure spray gun, which will, of course, need an air supply. You can rent these things, or possibly borrow one from the local garage or car repairers. Hang up the components somewhere dry and dust-free and apply numerous coats of well-thinned paint. Check with the makers about the correct thinner for their paint.

Failing a proper spray set, at least one specialist railway paint supplier produces aerosol cans which work quite well on small areas, though it is not so easy to get even coverage of large flat surfaces. A final choice is brushing which can also work well, given a decent-quality brush and a systematic approach. Cover the pieces with a cardboard box or similar whilst the paint dries, to stop rust and insects wreaking havoc. Whatever the finish used, allow it to harden thoroughly, several days at least, before attempting assembly.

Many full size locomotives, particularly the express ones, were extensively decorated with lettering and lining, and clearly we will want to reproduce this on the model if we can. Hand-lettering is a job for an expert, if you can get hold of one, but transfers make a good alternative and the trade produce a wide selection. Transfers will need a coat of varnish over the top to protect them, so get something compatible with the paint used. Lining also can be done using the transfers made for the job or, for a touch of class, have a go at hand lining yourself; it is not so difficult once you know the tricks, which I will endeavour to explain.

A good tool for the job is a spring-bow pen, as used by draughtsmen before the advent of the ubiquitous Rotring pens. This consists of two narrow, pointed, leaf-shaped blades linked by a screw to adjust the gap between them, and hence the width of line drawn. Use good quality gloss paint and dilute it down 50/50

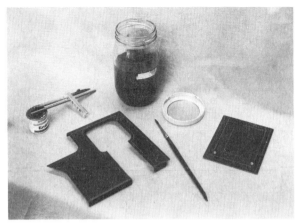

The set-up for lining out a cab side for 'Rob-Roy'. Note the distance leg fitted to the spring-bow pen.

with thinners, or use PVA water soluble artists' paint and varnish over it when dry. To load the pen, you do not just dip it into the paint, you use a suitable small paint brush and 'wipe' this into the gap between the pen blades. The pen is then run along the job until it runs out of paint, say about five or six inches of line, after which it must be cleaned. This involves another paint brush and a jar of solvent, such as acetone (nail varnish remover) which dries out very quickly, ready for the next load of paint. Sometimes you may find it best to start a line with a dot of paint applied with a toothpick. All this sounds tedious, but it really is not. One soon develops a rhythm and the results are delightful. Practise for a while on scrap material and you will soon be amazing other locomen at the model club, who will ask you to line out their engines too! Some people use the needle from a hypodermic syringe for lining and obtain excellent results.

Clearly, the pen needs some suitable guidance and one method is to use the edge of the job, fitting the pen with some form of edge follower, like odd-leg callipers. Alternatively, cut out a template from thin alloy, glue on some cardboard to act as a spacer, and carefully work around this. True, the job will require a fairly steady hand, plus a degree of patience and care, but it is not the mysterious black art that some people would have you believe.

Time spent on giving your model locomotive a good finish is well spent. It is, after all, the part that the world sees, and whilst paint can only cover a certain number of sins, a nice paint job can turn a toy into a model worthy of praise.

Well, that just about covers everything in the way of construction. We have discussed every aspect from main frames to paint, and I hope you now have some understanding of the sort of work involved in a model steam locomotive. The description has necessarily been general, covering no particular design, but the intention has been to suggest ways of tackling jobs and to expose some myths with the aim of helping a complete beginner not just to start an engine, but to finish it, to finish it completely, and to finish it well. So now, it's off to the track to get steam up and get the trains running. The next chapter applies whether the loco is that being steamed in bare metal or that being run first after full completion.

Chapter 10

On to the rails

Raising steam and driving

Raising steam from cold on full-size steam locomotives was a lengthy, but pretty simple job. Essentially, a fire was lit in the firebox, using wood initially, and, apart from throwing in a few shovels full of coal from time to time, it was just a question of waiting 24 hours or so. Natural draught was sufficient to keep the fire going, and the boiler was brought up to pressure gradually, which was good for the structure. Woe betide the engineman who left the loco in gear and with the regulator cracked at this time, because as pressure gradually rose, the engine would quietly drive itself away! The exhaust steam pulled the fire, pressure increased and all round embarrassment would ensue. The old steam railways were pretty dangerous places to work at times.

Our size of engine is too small for natural draught to draw the fire, besides which, no one wants to stand around for several hours whilst raising steam. The answer is that some form of artificial draught has to be provided until the boiler has reached, perhaps, 20 psi. After this the loco's own steam blower will take her on up to working pressure at 80 psi or so. Some clubs provide compressed air at the steaming bays, in which case a simple ejector can be used. This consists of a piece of tube around 6 in to 9 in long and which is a push-fit into the loco funnel. It carries a simple jet near the bottom end, pointing towards the top, and creates a vacuum in the same way as the loco blower. Most people, however, use an electric blower of some sort, usually designed to run off a car battery. All sorts of weird and wonderful contrivances made from biscuit tins and the like are seen. It is a question of pressing into service whatever comes to hand, although castings are now commercially available to make a rather neat little electric blower. Bear in mind that the thing will get quite hot in service and that the motor may object to this treatment!

Having prepared some sort of blower, it is necessary to prepare the loco. First of all we need some water, and the tap is not the best place to obtain it. Unless you live in an area with very soft water, the best plan is to use rain water, and most clubs collect from the clubhouse drainpipe for the purpose. Obviously, hard water scales up the boiler progressively, initially reducing the steaming rate and ultimately causing failure of the boiler by overheating, since the chalk acts as an insulator and stops the water dispersing the heat of the fire. Even when using pretty clean water there is a certain amount of deposition, and since most of the boiling is done around the firebox, it follows that most of the deposits fall on to the foundation ring. This accounts for its alternative name in traction engine parlance, the mud ring. The blowdown valves are sited here so that crud can be blown out when the valves are opened. Some full size locos had continuous blowdown in an attempt to deal with the scale problem.

Fill up the tender or tanks with approved water and fill the boiler, either by removing a safety valve and pouring it in, or by employing the nearest enthusiastic small boy to waggle the hand pump lever until the water level is just below the top nut of the water gauge. Fill the lubricator with steam oil and, if you can, give it a few turns by hand as a primer. Go around the rest of the loco with a can of car engine oil, fitted with a very long spout. If it moves, lubricate it; axleboxes, axles,

Right *Getting up steam on a 5 in-gauge 0-8-0, LBSC's 'Netta' design.*

Left *A 'Simplex' on the steaming bay. A good shot of the electric blower used to get things going.*

Bottom left *Pause for a chat whilst steam raising. This is the model loco hobby at its best.*

motion work, pump ram and so on.

At last the fire can be lit, starting with suitably sized pieces of wood, soaked in paraffin. Put in a layer of these on top of the grate, then place one on your shovel (see next section), apply a match to it, toss it into the firebox and start the blower going. Right at the start you will need to keep the draught low, or the fire will blow out. Once the wood has taken, pile in some more, up to about the level of the firehole, close the door and give it a couple of minutes, during which time you can check that the regulator and blower are closed and that she is in mid-gear. Small engines can also drive themselves away! The prudent types will also pull on the safety valve stems to make sure they are not jammed in any way, if the design so allows. Now start adding some small lumps of coal over the entire area of the fire, a little at a time to take over from the wood. Hopefully, the club will supply the fuel for the job. Welsh steam coal is ideal, but rare; anthracite is a good substitute. Whichever fuel, it should be broken up and sieved into half or three quarter inch cubes, approximately. Charcoal has been used as fuel, being very clean burning, but tends to burn through too quickly in a loco. Gas firing is occasionally used in models and is fine, but hardly scale British practice.

By now, the loco will probably be showing some pressure on the gauge and at around about 20–30 psi you can remove the external blower, with care because it is hot, and open up the loco blower valve. You will not be surprised to find that this also is hot, which is why many model loco drivers wear a leather glove. Keep on adding coal over the whole area of the fire, being careful to let the front end of the firebox get its share, particularly if it is an engine like, say, a 'B1', with a very long, narrow firebox. Keep the firehole door closed in between coaling. Keep an eye on the pressure gauge now as she comes up to the designed working pressure, and make sure that one safety valve blows at the designed pressure, and the second at a few pounds higher. These should have been roughly set under hydraulic pressure, but be ready with a suitable tool and gloved hands to adjust them again now. If your loco has Ross 'Pop' type safety valves, watch the audience jump when she suddenly blows off! Incidentally, steam raising should take only five or six minutes for the average engine.

Once you are happy with the valve setting, close down the blower to simmer, put the loco into full forward gear, open the drain cocks if she has them, and crack the regulator. Push the engine gently forward and perhaps give her a touch more regulator. The first lot of steam will condense in the cold steam chests and cylinders, and then will come spouting out of the funnel in a hot shower bath of oily water. Do not panic and let go of the engine! It will take quite a few revolutions of the wheels before it begins to clear and starts to pull. When she does, be ready to close the regulator down or the wheels may spin. Let the engine run like this until it stops spitting and starts to pull steadily, then you can hitch up the driving trolley and climb aboard. Check the fire, add coal if necessary and check the water level, which will certainly have fallen. If it is below half-glass, try out your injector, or use the hand pump for now. Replenish the water tank if necessary before running. When all is settled down to your satisfaction, put the loco into full forward gear again, open the draincocks, if not still open, and crack the regulator. Open up gently until the train starts to roll, and be prepared to shut down if the wheels slip. Once you are under way, close the drain cocks and bring the gear back a notch or two; then savour some of the sweet smell of success. If you have got this far, you deserve it!

Do not relax yet, though, because a locomotive and train need constant attention under way. Basically, there are two things that you can do wrong that could have serious consequences when driving a steam loco, full size or model. Firstly, you can go crashing into the back of another train, or bufferstops, or go into bends or points too fast. The answer here is plain common sense, obey the signals if there are any, and keep a good lookout. This may sound easy, but a steam loco needs looking after, since you are both driver and fireman, and it is very easy to forget about the track ahead.

The second disaster area is to let the water level in the boiler get too low. The crown of the firebox will then dry out, overheat, and fail, possibly explosively, which is unpleasant. Full size engines, and some models, have fusible plugs to prevent this happening. These consist of plugs in the firebox crown retained with soft solder which melts if the water level drops too low, thus releasing the boiler pressure. However, for the most part it is up to the driver to keep the kettle topped up at all times. Small water gauges are not easy to read and sometimes call for some

Above *A vital job, going round with the oil can whilst steam is being raised.*

Below *Boxing Day steam-up on a ground level track. Plenty of steam to be seen in the cold air!*

judgement in interpretation. Fast boiling action causes bubbles in the column and this is why water gauges often have blowdown valves, to clear these bubbles. If in doubt, stop and gently work the engine to and fro. If the level in the glass rises and falls in sympathy, you have a true reading. Most beginners' locos have axle pumps, and by adjustment of the by-pass valve it is possible to maintain water level fairly steadily, with some practice. Those with injectors can experiment with these, and if all else fails, there is the hand pump. In case of total panic, pull the pin and dump the fire.

Adding cold water to a boiler naturally tends to reduce the steaming rate, as does coaling the fire, and this is where a certain degree of skill is required. Bring in too much water and the pressure will drop away all of a sudden, forcing you to stop the train and open up the blower until pressure is higher. Unless the train is very heavy, 30 or 40 psi will probably be enough, because once the exhaust gets going, the fire will draw readily. Open up as far as wheel slip and prudence will allow, and the fire should recover from black in a few yards of running, to become a near-white glowing mass. The time to put in water and to add coal to the fire is when the safety valve is blowing and/or when descending a bank. Locomotive men in real life knew just how to mortgage the thermal reservoir of the boiler on an up grade, and to repay the debt on the down grades, where the demand for steam was less. This is something the good model driver will also acquire in time. Fire little and often, and monitor the water situation constantly. Aim to drive steadily and evenly as if the railways inspector were with you all the time.

Yet another thing to watch for, particularly on the first few laps of a new engine, is lubrication. If your face becomes spattered with black, oily spots and the funnel has an oily rim, then all is well. If no oil shows, and the engine begins to sound hoarse and tinny, then the lubricator needs some sort of attention. Lift the lid and if there is water on top, the check valves leak and require some improvement. This will entail cleaning out the oil line to prevent strange lock problems with the mixture of oil and water. The whole business of steaming and learning to drive are made a lot easier with the help of someone with experience, so do not be afraid to seek help at this time.

After the run, bring the engine back to the steaming bay and drop the fire down, not on to your feet, then open up the blowdown valve or valves to empty the boiler. This will actually take several minutes, or longer if the boiler was full at the end of the run, but once empty, the hot boiler will dry out quickly and no corrosion will occur in storage. Leave the regulator and all the wheel valves open in storage so that they do not jam. Incidentally, if the check valves or pump valves should stick in storage, a kettle of hot water will often sort them out for the next run.

The next job is tube sweeping, so put a rag over the front plate of the loco and open up the smokebox and firehole doors. The trade supply special bronze or bristle brushes for tube sweeping, and one of these is carefully pushed through each tube in turn. You may need a torch to see what is happening in a sooty black smokebox. The soot and char in the smokebox can then be cleaned out, the rim cleaned and the door closed. Replace the grate and ashpan when cold, disconnect and drain the tender if you have one — tank engines will probably have to be inverted to drain them out dry. Finally, clean the whole engine down with clean rag or kitchen paper and then oil the bright steel to preserve it. It is also worth checking round a new engine at this stage to make sure that none of the glands have begun to unscrew themselves, and that all the other pins, bushes, nuts, bolts and the like are all present and correct. Let the engine cool fully before boxing for storage or condensation corrosion may be aggravated.

Proper cleaning up after a run does take time, but it is worth the effort to preserve an engine in the best possible condition. In this way she may run for many years before overhaul, reboring and rebushing become necessary. As I have indicated, your engine will probably outlast you!

Tools, trolleys and tracks
Only a few basic tools are required for running a model steam loco, apart from the blower device. One is a decent shovel which will pass through the firehole and reach to the other end of the grate, so that coal can be distributed to all parts evenly. This can be easily made from sheet steel and brazed on to a length of rod for a handle. Next comes a fire poking or pricking device for raking through the fire to drop down ash and free a choked grate. The two can be made as a matching pair, and oil-blacked. As mentioned, an oil can with a very long spout, made out of a foot of $\frac{1}{8}$ in thick wall copper tube, is handy. This should reach parts of the engine that other oil cans cannot, if you see what I mean. Flue brushes are commercial items, and of course, you will need a handle for the hand pump. Pliers may be needed for the ashpan dump-pin, and spanners for checking unions, etc. An engineman's greasetop hat is, of course, the correct headgear for the occasion, and these come from railway enthusiasts' shops, making great Christmas presents.

Like blowers, riding trolleys come in all shapes and sizes, from beautifully engineered carriages with two ball-raced bogies, to nailed up wooden four-wheel horrors. The problem is, I suspect, that having spent a lot of time on the loco, people rush together a trolley to ride on, then go on to the next project. LBSC produced some designs for riding cars and drawings and castings are available for these from the model supply people. You should, however, consult with the local clubs if you plan to run on their tracks. For one thing, it may be easier to use club trolleys, or bogies as they are also known, and not to bother with making anything at all. For another, the nature of the track will dictate the design of riding car, and the club may also have particular preferences or requirements which you will need to know about.

In the past, raised tracks were almost universal, providing a nice comfortable ride, particularly behind the smaller gauge engines. Nowadays, the bigger gauges have become popular and more ground level tracks have appeared, catering for $7\frac{1}{4}$ in gauge. My local club is blessed with both types of track, the raised one for $3\frac{1}{2}$ in and 5 in gauges, and the ground level for $3\frac{1}{2}$ in, 5 in and $7\frac{1}{4}$ in gauges.

I have a riding car of only $3\frac{1}{2}$ in gauge which has proved adequately stable if used sensibly, but in the main trolleys should be of as wide a gauge as possible for maximum safety. This leads to the slightly anomalous sight of $3\frac{1}{2}$ in-gauge engines hauling $7\frac{1}{4}$ in-gauge bogies, which is no great problem except in that the bogie

A 7¹/₄ in-gauge drivers's trolley with space for the inevitable accumulation of tools and junk.

must have an offset drawbar to cater for the displaced centre lines of engine and train.

The simplest driving trolley is just a four-wheeled truck with space for the driver and perhaps extra coal and water, and maybe tools. The wheels may be rigidly fixed to the chassis, mounted to rest on rubber door stops, or fully spring suspended, but either way, ball or roller races are essential to cut down resistance. Starting away is the most difficult time for the engine, and something like 150 lb resting on four plain bearings will require a pretty massive draw-bar pull to get rolling. Some form of upholstery will make the ride more pleasant, particularly if you envisage long spells of passenger hauling on club open days. Footrests are absolutely essential for raised track running and are most comfortable when set slightly ahead of the sitting position, but beware of tipping the trolley forwards like this. Good brakes are important on a driver's car, and they are fairly easy to fit on simple, four-wheel trucks. Crude, but quite effective, is a plate which is pressed down on to the rails by a lever. Rather better is some arrangement to pull rubber bicycle brake blocks against each wheel. The best way to connect the trolley to the loco, and to any following trolleys, is not by chain, but by a rigid bar. This is retained by a steel pin through a double bracket on loco and trolley, and the pin should have some positive device to prevent it jumping free.

Rather better than a rigid, four-wheel driver's truck is a longer car mounted on two four-wheel bogies; however, this will, of course, involve considerably more work. As mentioned, castings and drawings for these are available commercially, but a visit to one or two model tracks will show you the sort of thing that can be done. Once again, ball or roller races are desirable, as are brakes, but these are more of a problem since the bogies must swing free. Ideally, the two wheels on each axle should be free to rotate independently so that they have some

A very nice 5 in-gauge riding trolley for use on raised tracks. Note the brakes and multiple coupling to cope with different engines.

'differential' on corners. This helps to cut down rolling resistance to a minimum. My own 3½ in-gauge trolley has ball races throughout, with one loose wheel on each axle, and runs very freely indeed. This was necessary since it was designed to run behind a very small engine with only one or two passengers. It therefore has no brakes and relies on the driver's feet in emergencies! A very nice design of trolley was described in *Model Engineer* No 3753, May 1985.

In the larger gauges, and where heavier trains are handled, then the riding cars may be fitted with properly engineered air or vacuum brake systems. This is vital for operations on the scale of the Dobwalls 7¼ in-gauge railway, but of course this is a full-scale commercial venture, with trains a good deal larger than the average beginner will encounter!

Model tracks also come in a variety of forms and materials. First there is the division between raised and ground level, then the choice of steel or alloy rail, and so on. The beginner may choose to build some track of his own, but unless totally isolated from the rest of the world by geography, the best bet will be to run on the local club track. If there is no club near you, then I suggest you start one!

Raised tracks generally stand about 18 in above ground level on posts and beams of wood, concrete or metal. The track itself usually has wooden sleepers, grooved to carry the rail head for two or three gauges, say, 2½ in, 3½ in and 5 in. The rail is very often mild steel bar of about 1 in by ⅛ in section, which is an excellent material, being cheap and also providing good adhesion. This latter point is quite significant, since the hauling power of a model loco, much like full size, is often limited by adhesion of the driving wheels. Steel, of course, rusts continually and thus provides a roughened surface for the wheels to bite on. I have encountered brass rail at one club track but, although this gave beautifully free running for very small engines, cost would normally prohibit the use of brass.

Left *The bogie and brake gear of the trolley shown on page 205.*

Right *A 5 in-gauge bogie for ground level operation.*

Bottom right *Bogie and brake gear on this one. The foot boards show evidence of ballast damage, caused by rabbits digging up the home track bed!*

Extruded aluminium alloy is the main alternative to steel and this is available in scale-ish, flat-bottom rail section, but has the disadvantage of very poor adhesion, particularly if any oil is deposited on the track. It is often used for ground level tracks, retained on to pre-drilled wooden sleepers by galvanized screws. Ground level tracks look much better, being more like the real thing, but tend to be less comfortable for driving. One is continually crouched down over the engine, particularly if it is a small one, creating problems for the less fit. Ground level track is, however, cheaper to build per foot and requires less work than raised types, whilst it also allows point work and crossings which are otherwise not practical. One other positive point is that a derailed loco at ground level is less liable to damage than one which falls from a raised track. On the other hand, ground level tracks tend to need constant attention to levelling and alignment, while raised tracks remain more stable.

Model club railway tracks have improved considerably in the last ten or fifteen years, when people have had more time, money and enthusiasm to work on this sort of project. In the past a simple oval was often all that was achieved. Now, scenic routes wind through woods, across embankments and bridges, through cuttings and tunnels. There are long straights and tortuous curves, level sections and banks, laid out to challenge driver and loco together. Signalling is not just an embellishment on such tracks, but a safety requirement. Stations and steaming bays with various services, electricity and water are provided and, all in all, the modern club track can provide much entertainment for the whole family. The photographs will give you some idea of the sort of thing that can be achieved by banding together to construct a layout.

Above *Model railway in silvan surroundings. This is the Cambridge club track showing the steaming bays and ground level line.*

Below *A shot of the raised track at Cambridge club, showing the method of construction.*

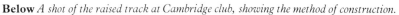

Above *This is the station at Cambridge in GWR colours — not what you might think, GWR stands for Grantchester Woodland Railway!*

Below *Raised track point work involves a traversing section like this one.*

Above *Ground level multi-gauge track for 3¹/₂ in, 5 and 7¹/₄ in locos. The rail head is commercially available aluminium extrusion. Note the gaps to allow for expansion as per full size.*

Below *This shows both the raised and ground level tracks side by side at Cambridge. Multi-gauge point work has to be very carefully designed to accommodate the wheel standards of the various gauges. This point was designed by a professor of mathematics!*

The model clubs, competitions and exhibitions

I have already made frequent reference to the model engineering clubs which exist in most towns throughout this country, because they are very important to the model locomotive enthusiast. Not only do they provide the practical items like the track and sometimes a workshop, but they constitute a pool of experience and a focus for interest in the hobby. Most embrace not only steam locomotives, which constitute probably the biggest branch of model engineering, but also the other aspects, such as traction engines, stationary engines, petrol and similar engines, and often clock making too.

The typical club has, perhaps, eighty members of which probably only twenty are really active, and a clubhouse-cum-workshop on the track site. There will be regular, informal gatherings at weekends, and maybe one evening in the week, to run models, drink tea and chat. Formal meetings may be held monthly, at which there could be a guest speaker, a demonstration of some kind, film show, slide show, or whatever. Most clubs hold competitions for models, both in the static condition and under working conditions, and these may be associated with local exhibitions of the model builder's art. This kind of show may well link up with the model aircraft and boat enthusiasts in the area. Invitation rallies are a prominent feature of club timetables and involve throwing open the doors to other clubs to come and run their models on your track. Fund raising is also an important activity, both for outside charities and also to pay club bills for rent, rates, electricity and so forth. Public open days are hard work but fun, and can raise considerable sums of money. Some clubs have portable tracks which can be loaded on to a trailer and taken to fetes or similar events, to give rides to the children. For a list of model clubs, see Appendix 2.

On a national scale, there are two major model engineering exhibitions each year, and there is the annual International Model Locomotive Efficiency Competition (IMLEC). The Wembley exhibition, held just after Christmas at the Wembley Conference Centre, is the premier event, dating back over fifty years. Hundreds of models of all shapes and sizes are on show, and some very fine locomotives indeed can be seen at close range. In addition, many of the trade suppliers will be present and this is a good opportunity to stock up on materials and to see what is available. The Midlands Exhibition is a similar sort of event held in November in the Birmingham area. A portable track is very often laid out at these events and there will be locos in steam, hauling passengers. Exhibitions of this sort are a great experience and a source of inspiration for the beginner and expert alike.

The IMLEC event is held each year at a different club track, over a weekend, and attracts a wide field of contestants. The competition is for efficiency; that is, the coal used during a standard half hour run is measured, the work done is assessed by a dynamometer car, and a figure is computed for the efficiency of the loco. The winner has the highest percentage efficiency, a figure usually around 2 per cent. This is an extreme test for the engine and the skill of the driver, who must choose how much load to pull, and then keep this train steadily rolling for thirty minutes whilst using the minimum quantity of coal. Again, this is an event worth

Above *Typical scene at the steaming bays of a model club on invitation day.*

Below *Ascot Model Locomotive Club probably has the finest multi-gauge track in the country. This shows part of the station, note the signals, signal box and all-round neatness. (Steve Stephens.)*

attending because the finest engines and drivers can be seen at work.

This gives the newcomer to model steam locomotives some idea of the sort of social background to the hobby. Thus, I think I have covered all the aspects of model steam locomotives, from how they work, how they are built, to the equipment needed, and what to do when you have your loco. I hope that many questions will have been answered and that the beginner can embark on buying or building a loco knowing what is involved, where to obtain help, materials, tools and so forth. In short, I hope to have explained a lot of the peripheral detail that many of the specialist books either skate over briefly or leave out entirely. If this book spurs you through to complete a locomotive, then seek me out at any track events you go to, and introduce yourself, because I look forward very much to meeting you and your model steam locomotive!

Appendix 1

Members of the Southern Federation of Model Engineering Societies

Andover and District Model Engineering Society, Berks.
Ascot Locomotive Club, Berks.
Ashstead Miniature Steam Railway Society, Surrey.
Banbury and District Model Engineering Society, Oxon.
Barnsley Society of Model Engineers, South Yorks.
Basingstoke and District Model Engineering Society, Berks.
Bedford Model Engineering Society, Beds.
Bexhill and Hastings Society of Model Engineers, East Sussex.
Bishops Stortford Model Engineering Society, Herts.
Bournemouth and District Society of Model Engineers, Dorset.
Bracknell and District Model Railway Society, Berks.
Brandon and District Society of Model Engineers, Norfolk.
Brighton and Hove Society of Miniature Locomotive Engineers, Sussex.
Bristol Society of Model and Experimental Engineers, Avon.
Bromsgrove Society of Model Engineers, West Midlands.
Cambridge Model Engineering Society, Cambs.
Cannock Chase Model Engineering Society, Staffs.
Canterbury and District Model Engineering Society, Kent.
Canvey Railway and Model Engineering Club, Essex.
Cheddar Valley Society of Model Engineers, Avon.
Chelmsford Society of Model Engineers, Essex.
Cheltenham Society of Model Engineers, Glos.
Chichester and District Society of Model Engineers, West Sussex.
Chingford and District Model Engineering Club, Essex.
Colchester Society of Model and Experimental Engineers, Essex.
Crawley Model Engineers, West Sussex.
Croydon Society of Model Engineers, Surrey.
Didcot and District Model Engineering Society, Oxon.
Doncaster Society of Model Engineers, South Yorks.
East Herts Miniature Railway, Herts.
East Sussex Model Engineers, East Sussex.
Edinburgh Society of Model Engineers, Lothian.
Erith Live Steamers, Kent.

Esk Valley Model Engineering Society, Lothian.
Exeter and District Model Engineering Society, Devon.
Fareham and District Society of Model Engineers, Hants.
Farnham and District Model Railway Club, Surrey.
Glenrothes and District Model Engineering Society, Fife.
Gravesend Model Marine and Engineering Society, Kent.
Guildford Model Engineering Society, Surrey.
Halesowen and Leasones Model Engineering Society, West Midlands.
Halesworth and District Model Engineering Society, Suffolk.
Halsey School Miniature Railway, Hemel Hempstead, Herts.
Harlington Locomotive Society, London.
Harrow and Wembley Society of Model Engineers, London.
Hatfield and District Society of Model Engineers, Herts.
Hereford Society of Model Engineers, Hereford.
Herne Bay Model Railway and Engineering Club, Kent.
High Wycombe Model Engineering Club, Bucks.
Hitchin and District Model Engineering Club, Beds.
Hull Society of Model Engineers Ltd, North Humberside.
Ickenham and District Society of Model Engineers, London.
Ilford and West Essex Model Railway Club, Essex.
Ipswich Model Engineering Society, Suffolk.
Isle of Wight Model Engineering Society, Isle of Wight.
James Burns International Model Engineering Society, Esher, Surrey.
Kent Model Engineering Society, Kent.
Kidlington Society of Model Engineers, Oxford, Oxon.
Kings Lynn and District Society of Model Engineers, Norfolk.
Kinver and West Midlands Society of Model Engineers Ltd, West Midlands.
Lea Valley Railway Club, London.
Leeds Society of Model and Experimental Engineers, West Yorks.
Lindsey Model Society, South Humberside.
Liphook Modellers' Club, Hants.
Liskeard Model Society, Cornwall.
Littlebrook Sports and Social Club Model Engineering Section, Kent.
Luton and District Model Engineering Society, Herts.
Maidstone Model Engineering Society, Kent.
Malden and District Society of Model Engineers Ltd, Surrey.
Merthyr Tydfil Technical College Model Engineering Society, Mid Glamorgan.
Merton Society of Model Engineers, Surrey.
Milton Keynes Model Society, Bucks.
Newport Model Engineering Society, Gwent.
North Cornwall Model Society, Cornwall.
North Devon Society of Model Engineers, North Devon.
North London Society of Model Engineers, Herts.
Northolt Model Railway Club, London.
North Wilts Model Engineering Society, Wilts.

Norwich and District Society of Model Engineers Ltd, Norfolk.
Perranporth and District Model Engineering Society Ltd, Cornwall.
Peterborough Model Engineering Society, Cambs.
Pinewood Miniature Railway Society, Berks.
Plymouth Miniature Steam Locomotive Society, Devon.
Polegate and District Model Engineering Club, East Sussex.
Portsmouth Model Engineering Society, Hants.
Ramsgate and District Model Engineering Club, Kent.
Reading Society of Model Engineers, Berks.
Rivet Works Sports and Social Club Model Engineering Section, Aylesbury, Bucks.
Romford Model Engineering Club, Essex.
Romney Marsh Model Engineering Society, Kent.
Royal Navy Aircraft Yard, Fleetlands Model Engineering Society, Gosport, Hants.
Royston and District Model Engineering Society, Herts.
Rugby Model Engineering Society, Warwicks.
Ryedale Society of Model Engineers, Yorks.
Saffron Walden and District Model Engineering Society, Essex.
St Albans and District Model Engineering Society, Beds.
Scunthorpe Society of Model Engineers, South Humberside.
Sheppey Miniature Engineering and Model Society, Kent.
Southampton and District Society of Model Engineers, Hants.
South Birmingham Model Engineering Society, West Midlands.
South Durham Society of Model Engineers, County Durham.
South East Essex Railway Society, Essex.
Staines Society of Model Engineers, London.
Stamford Model Engineering Society, Lincs.
Standard Telephone and Cables (Paignton) A & SC Model and Engineering Society, Devon.
Surrey Society of Model Engineers, Surrey.
Sussex Miniature Locomotive Society Ltd, Sussex.
Sutton Model Engineering Club Ltd, Surrey.
Sutton Coldfield and North Birmingham Model Engineering Society, West Midlands.
Swansea Society of Model Engineers, West Glamorgan.
Taunton Model Engineers, Somerset.
The Great Cockcrow Railway, London.
Tiverton and Blundells Model Engineering Society, Devon.
Tonbridge Model Engineering Society, Kent.
Uckfield Model Railway Club, East Sussex.
Vale of Aylesbury Model Engineering Society, Bucks.
Vauxhall Motors Model and Engineering Section, Luton, Beds.
Welling and District Model Engineering Society, Kent.
Welwyn Garden City Society of Model Engineers, Herts.

Westland Model Engineering Society, Somerset.
Weston-super-Mare and West Huntspill Live Steam Society, Avon.
West Wiltshire Society of Model Engineers, Wilts.
Weymouth and District Model Engineering Society, Dorset.
Whitchurch (Cardiff) and District Model Engineering Society, South Glamorgan.
White Rose Model Engineering Society, Leeds, West Yorks.
Willesden and West London Society of Model Engineers, Bucks.
Wimborne and District Society of Model Engineers, Dorset.
Winchester Model and Engineering Society, Hants.
Witney and West Oxfordshire Society of Model Engineers, Oxon.
Woking Grange Miniature Railway, Hants.
Worthing and District Society of Model Engineers, West Sussex.
Yeovil College and District Model Engineering Society, Somerset.
York City and District Society of Model Engineers, North Yorks.

Appendix 2

Societies affiliated to the Northern Association of Model Engineers

Blackburn Model Engineering Society, Lancs.
Bradford Model Engineering Society, West Yorks.
Burton on Trent Model Engineering Society, Staffs.
Carlisle and District Model Engineering Society, Cumbria.
Chesterfield and District Model Engineering Society, Derbys.
Cleveland Association of Model Engineers, Darlington, Durham.
Crewe Model Engineering Society, Cheshire.
Derby Locomotive Works Society of Model Engineers, Derbys.
Erewash Valley Model Engineering Society, Peterborough, Cambs.
Furness Model Railway Club, Cumbria.
Fylde Society of Model Engineers, Lancs.
Glasgow Society of Model Engineers, Strathclyde.
Grimsby and Cleethorpes Model Engineering Society, Humberside.
Halton Miniature Railway Society, Runcorn, Cheshire.
Huddersfield Small Locomotive Society, West Yorks.
Keighley and District Model Engineering Society Ltd, West Yorks.
Lancaster and Morecambe Model Engineering Society, Lancs.
Leicester Society of Model Engineers, Leics.
Leyland Society of Model Engineers, Lancs.
Lincoln Model Engineering Society, Lincs.
Locomotive Model Society, Chester, Cheshire.
Macclesfield and District Model Engineering Society, Cheshire.
Manor Park Miniature Railway Co Ltd, Glossop, Derbys.
Marston Green Model Engineering Club, Birmingham, West Midlands.
Melton Mowbray and District Model Engineering Society, Leics.
Merseyside Live Steam and Model Engineers, Mosley Hill, Merseyside.
Mid-Cheshire Society of Model Engineers, Crewe, Cheshire.
Model Engineers Society Northern Ireland, Cultra, Co Down.
Mold Model Engineering Society, Clwyd.
Northampton Society of Model Engineers Ltd, Northants.
North Staffs Models Society, Newcastle under Lyme, Staffs.
North West Leicestershire Model Engineering Society, Coalville, Leics.
Ogwen Engineering Society, Bethesda, Gwynedd.

Phoenix Model Engineering Society, Telford, Salop.
Rochdale Society of Model and Experimental Engineers, Greater Manchester.
Sale Area Model Engineering Society, Greater Manchester.
Sheffield Society of Model and Experimental Engineers, South Yorks.
Shrewsbury and District Model Engineers Society, Salop.
Southport Model Engineering Club, Merseyside.
Spenborough Society of Model and Experimental Engineers, York, North Yorks.
Strathaven Model Society, Lanarks.
Stockport and District Society of Model Engineers, Greater Manchester.
Urmston and District Model Engineering Society Ltd, Greater Manchester.
Warrington and District Model Engineering Society, Cheshire.
West Cumbria Guild of Model Engineers, Cumbria.
Whitefield and District Model and Engineering Society, Greater Manchester.
Wigan and District Model Engineering Society, Greater Manchester.
Wirral Model Engineering Society, Merseyside.
Wrexham and District Society of Model Engineers, Clwyd.

Bibliography

Simple Locomotive Construction—Introducing LBSC's 'Tich', LBSC, Model and Allied Publications 1968.

The Model Steam Locomotive — A Complete Treatise on Design and Construction, Martin Evans, Argus 1983.

The Manual of Model Steam Locomotive Construction, Martin Evans, Model and Allied Publications 1968.

Shop, Shed and Road, LBSC, Model and Allied Publications 1969.

Model Steam Locomotives, Henry Greenly, Cassell and Company Limited 1922.

Model Locomotive Boilers, Martin Evans, Model and Allied Publications 1969.

Model Locomotive Valve Gears, Martin Evans, Argus 1981.

LBSC — His Life and Locomotives, Brian Hollingsworth, Croesor Junction Press 1983.

The Home Metal Worker, Johnathon Worthington (Series Editor), Orbis Publishing 1982.

Introducing Benchwork, Stan Bray, Patrick Stephens Limited 1986.

Introducing the Lathe, Stan Bray, Patrick Stephens Limited 1985.

Using the Small Lathe, L.C. Mason, Model and Allied Publications 1963.

The Amateur's Workshop, Ian Bradley, Argus 1984.

The Amateur's Lathe, Laurence Sparey, Argus 1984.

Model Engineer's Handbook, Tubal Cain, Argus 1984.

Hardening, Tempering and Heat Treatment, Tubal Cain, Argus 1984.

Milling Operations in the Lathe, Tubal Cain, Argus 1984.

Vertical Milling in the Home Workshop, Arnold Throp, Model and Allied Publications 1977.

Foundry Work for the Amateur, B. Terry Aspin, Model and Allied Publications 1954.

Index